DISSECTION OF THE
CAT

Steven W. Binkley

Illustrated by Sandra Shomaker
Consultant on Art and Anatomy
Kenneth W. Perkins, Ph.D.

Copyright ©1985
REX Educational Resources Company
2700 York Road
Burlington, North Carolina 27215

ISBN 0-89278-051-7
Library of Congress Catalog Card Number: 85-60676

CONTENTS

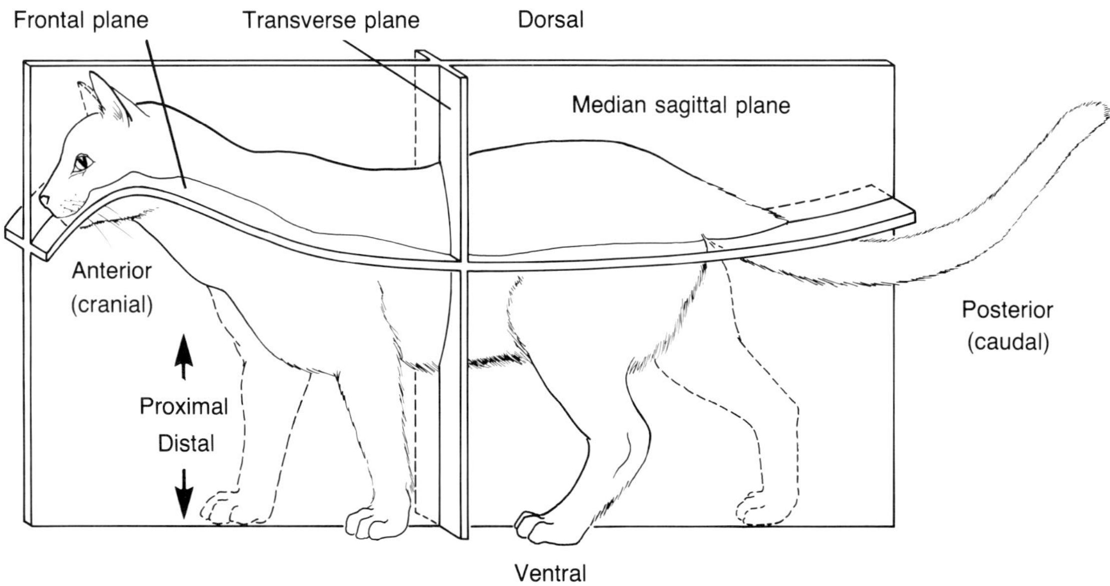

Figure 1. Body planes and directions.

INTRODUCTION

"...if you want to be a psychological novelist and write about human beings, the best thing you can do is keep a pair of cats."

Aldous Huxley

We made several assumptions in producing this dissection guide. First, we assume that this is the student's first dissection of an adult mammal; therefore, we present a general dissection as opposed to an exhaustive study. Another major assumption is that this should be a dissection guide only, and that information about the physiology, histology, embryology, etc. of the organs and systems dissected will be available from other sources. This led us to place the primary emphasis on the artwork as opposed to the text. Primarily, the text is used to give directions, explain terms, and to provide information helpful in interpreting what is seen. Anatomical terms are italicized at their first occurrence or when they are defined. Exceptions are made for some commonly known terms. We hope the use of detailed artwork will facilitate the location and identification of structures, decrease confusion, and encourage better understanding and dissection. Finally, we assume that the entire dissection will be done on one specimen (the use of two specimens is preferred, but economics may rule this out), supplemented by study of a disarticulated and an articulated cat skeleton. A triple-injected specimen is recommended.

"Normal" cat anatomy is presented throughout this dissection guide. Of course, there really are no "normal" cats. There are only individual cats with individual variations, and your specimen may show marked differences from our illustrations. For example, one cat dissected in the production of this guide had a posterior vena cava that formed two equal branches just anterior to the renal veins.

We end each chapter with a few questions and/or suggestions for short projects. Their intent is to stimulate discussion and curiosity, and to cause you to consider the wider implications of your study. We trust their inclusion will not limit your own questioning.

Why dissect a cat? Charts cannot give a complete idea of what an organ or organ system is like. The concentration and involvement in searching out structures aids learning and develops skills that may be of later use. A cat is large enough for its internal organs (which are much like our own) to be found easily and for its blood vessels to be traced, but not so large as to present special difficulty in storage and handling.

Laboratory Safety

While dissecting, observe these safety precautions. Wear safety glasses or regular eyeglasses to protect the eyes from possible splatters with preservative fluid. Contact lenses should be removed. If your eyes are especially sensitive to preservative fumes, close-fitting safety goggles may help. Rinsing the specimen will also be helpful. Latex gloves will protect the hands, and a lab coat will protect clothes and arms. Sharp instruments must be used with caution.

Orientation

It will often be necessary to refer to certain directions or positions relative to the cat's body or its organs. Figure 1 will help you understand the terms used. The figure shows a *left lateral* view of the cat. (Remember, left and right refer to the cat's left and right.)

Lateral — toward the side.

Medial — toward the midline.

Proximal — near or toward the point of reference (e.g. midline of body).

Distal — away from the point of reference (e.g. midline of body).

Dorsal — toward the back.

Ventral — toward the belly.

Anterior (cranial) — toward the head.

Posterior (caudal) — toward the tail.

In observing the positions and relationships of internal structures, it is useful to imagine the body as being sectioned along certain fixed planes of reference. There are three major planes.

Sagittal plane — divides the body or organ into left and right portions. A mid-sagittal section has equal right and left halves. Figure 1 shows a *median sagittal plane* that divides the body into equal right and left halves. A sagittal cut produces a *longitudinal section*.

Transverse plane — perpendicular to the long axis. All sections cut along transverse planes are *cross sections*.

Frontal plane — divides the body into dorsal and ventral portions.

Unit 1
EXTERNAL FEATURES

Spend a few moments examining the external anatomy of your cat. The following discussion covers the most obvious features.

Like most mammals, the cat is almost completely covered with a dense coat of hair. The primary function of mammalian hair is insulation, but there are interesting secondary functions as well. The cat has about its mouth, cheeks, and eyes stout, long hairs called *vibrissae*. These have a sensory function, allowing the cat to "feel" its way in darkness.

Other important sense organs located on or opening onto the head include the eyes, ears, and nose. The *external nares* (nostrils) are separated by a groove, the *philtrum,* which forms a cleft in the upper lip. In man the philtrum is normally closed and flat. The eyes face forward, giving the cat stereoscopic vision. Spread apart the upper and lower eyelids and locate, in the medial corner of the eye, the third eyelid or *nictitating membrane*. In life this is a transparent layer that helps protect and cleanse the eye. Mammals are the only animals that have *pinnae*, external ear structures. These pinnae protect the ear openings and reflect sound into the middle ear.

Mammalian digits (toes) are usually tipped with claws, hooves, or nails. Carnivores have claws, and the cat's claws are highly specialized so that they can be retracted. This protects them from being dulled by contact with the ground, yet they can be brought into play when needed. Notice that one claw on each foot is raised above the level of the rest. On the undersurface of each foot are epidermal thickenings called friction pads or *tori* (singular, torus). These cushion the step and give traction. There are seven tori on each forefoot and five on each hindfoot.

Mammals are animals with mammary glands. Cats have two rows of about five mammary glands each. Look for the nipples to the left and right of the ventral midline.

The cat's tail is well developed and often expresses the cat's mood. The *anus,* the posterior opening of the digestive system, is located just ventral to the base of the tail.

The external genitalia are the external sexual structures. Examine specimens of both sexes for these. The *urogenital aperture,* the opening of the female urogenital tract, is just ventral to the anus. At a similar location on the male is the *scrotum,* a sac containing the testes. Immediately anterior to the scrotum is the *prepuce,* a slight swelling into which the penis is normally retracted.

Questions and Activities

1. Thus far you have observed at least two features that are unique to mammals. List these.
2. List the body surface areas that are not covered by hair. Give possible reasons for the lack of hair in these areas.
3. List some external features of your specimen that are specializations for a predatory life.

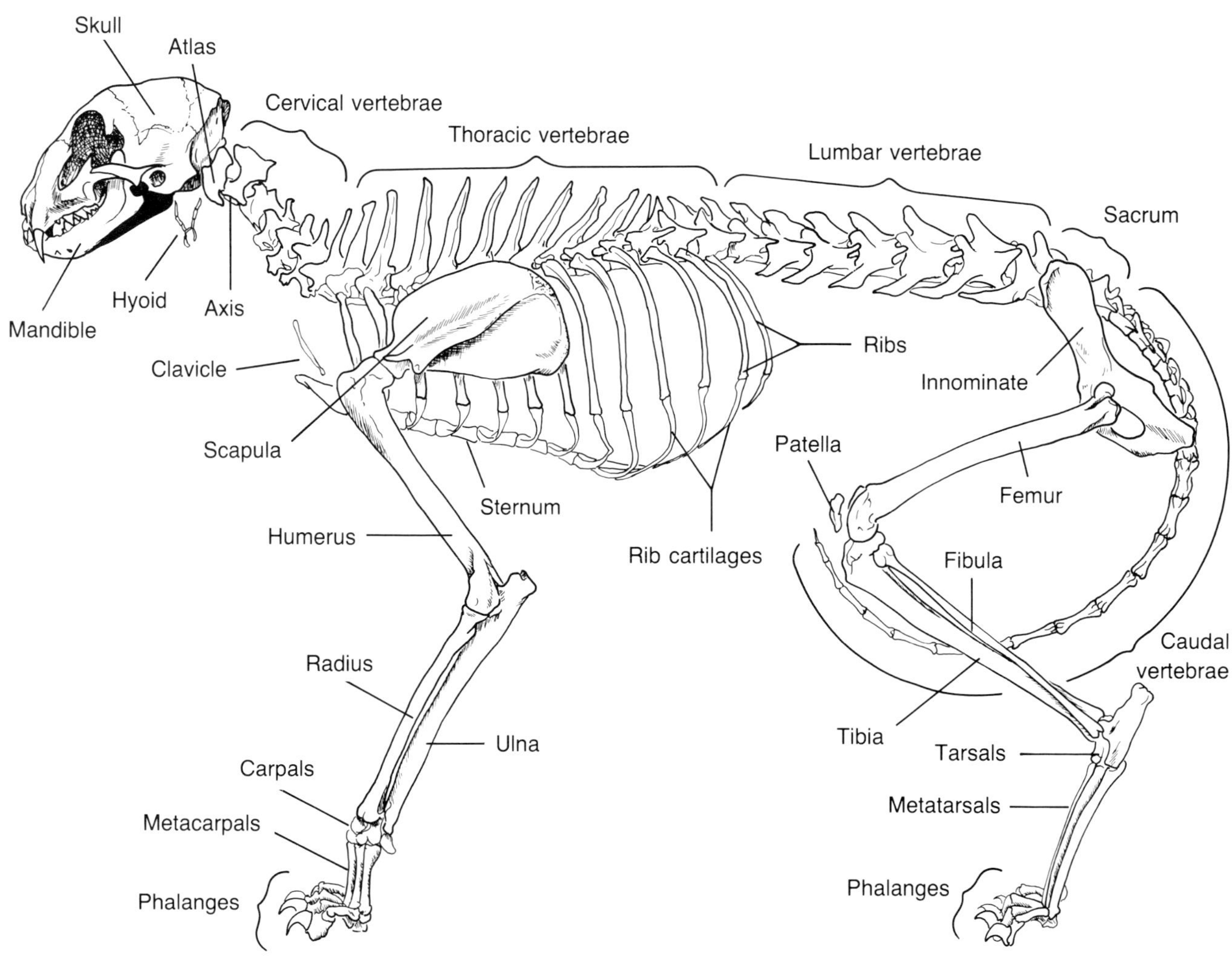

Figure 2. Skeleton.

Unit 2
THE SKELETON

The skeleton is the supporting framework of the body. It also serves for the attachment of major muscles and provides a system of levers by which muscular contractions are converted to bodily movements. A knowledge of the skeleton will help you understand basic body structure and the actions of the muscles.

Begin by observing a mounted cat skeleton. The skeleton has two major divisions: the *axial skeleton,* consisting of the skull, mandible, vertebral column, ribs, and sternum; and the *appendicular skeleton,* consisting of the pectoral girdle, forelimbs, pelvic girdle, and hind limbs. The animal's body is slung from a more-or-less horizontal backbone, and the backbone, in turn, is supported near either end by a pair of limbs.

Move the mandible up and down (if it is moveable) and notice how the upper and lower teeth mesh. Also notice that the posterior teeth (premolars and molars) slide past each other in a scissors-like action. Observe the movement of the mandible at the joint where it articulates with the skull.

Now look at the posterior end of the skull and locate the ridge and roughened area that served for the attachment of heavy neck muscles. These balance the head at the end of the vertebral column. The skull articulates with the vertebral column by means of two bony knobs, the *occipital condyles* (*condyle:* a rounded process used for articulation). These rest against the *atlas,* the most anterior of the vertebrae. Both the atlas and the vertebra with which it articulates (called the *axis*) are uniquely shaped and easily distinguished from other vertebrae and from each other. The vertebrae are divided into five groups, as shown in Figure 2. The *cervical vertebrae* are the most anterior seven. The next thirteen vertebrae articulate to ribs, and are called the *thoracic vertebrae.* Then come the seven *lumbar vertebrae.* The *sacrum* is formed by the fusion of the three sacral vertebrae. Posterior to the

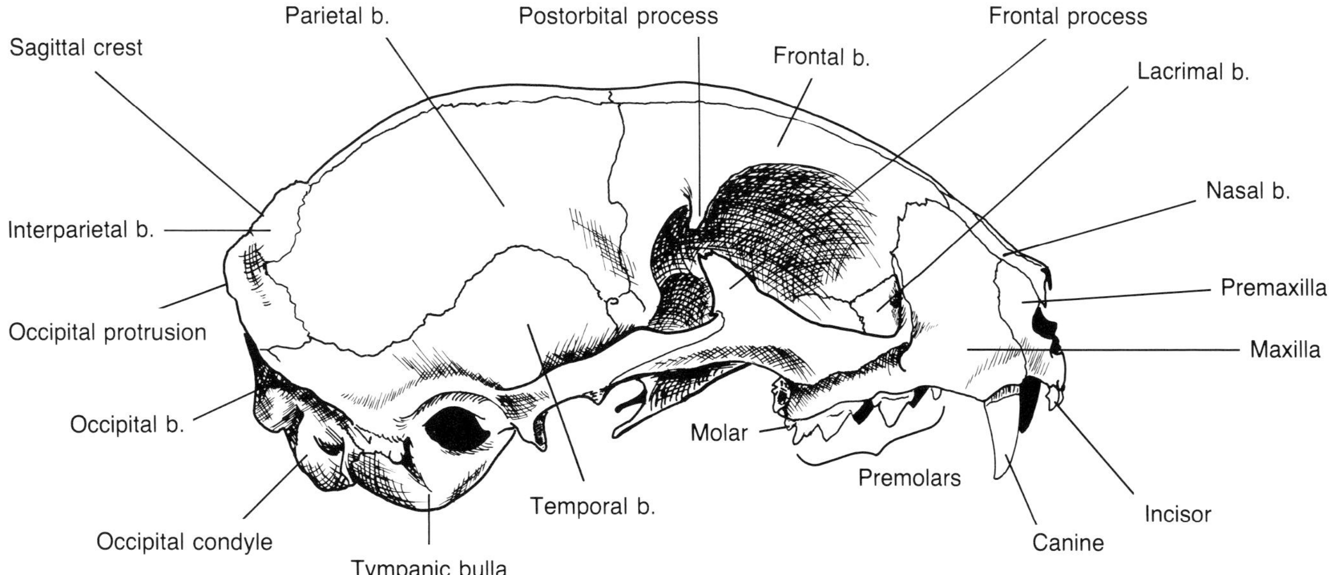

Figure 3. Skull (right lateral view).

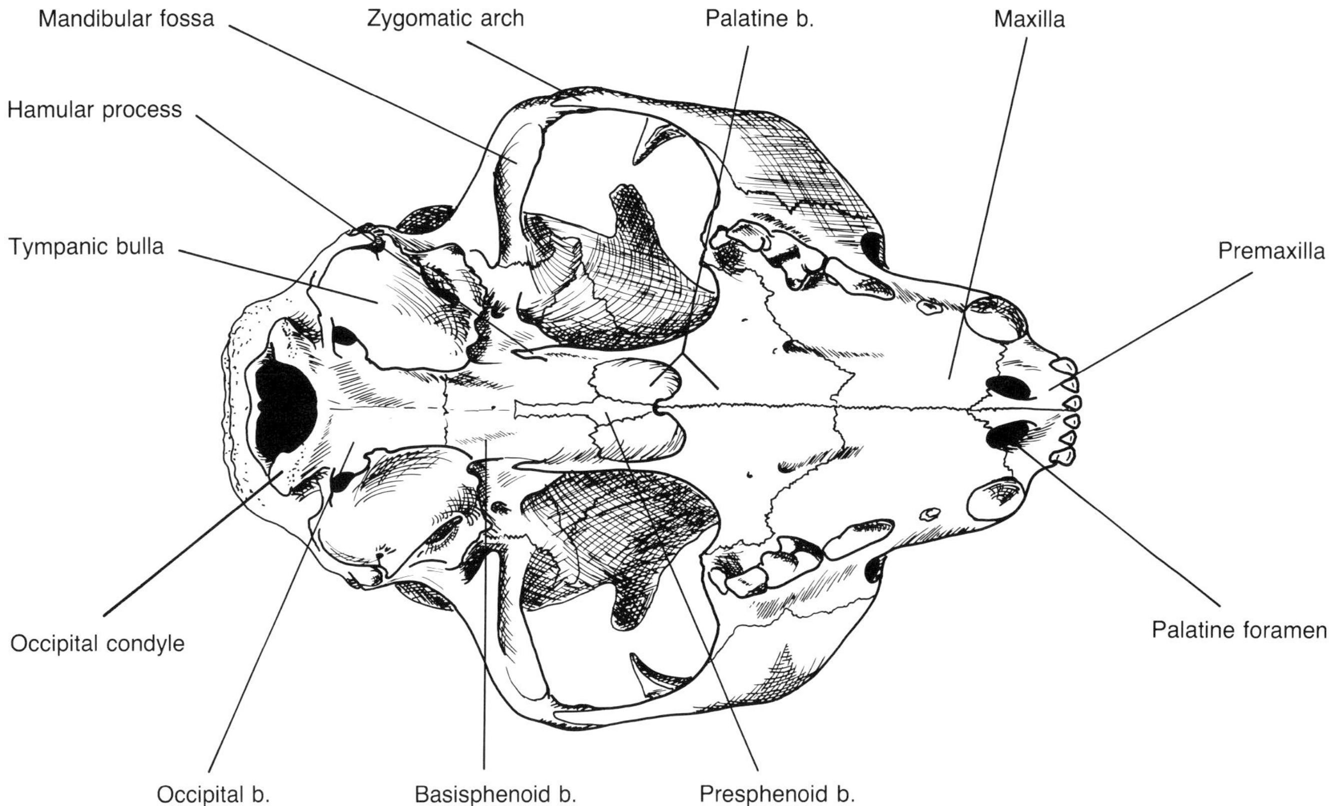

Figure 4. Skull (ventral view).

sacrum are the *caudal* or tail vertebrae. Spend some time becoming familiar with the different groups of vertebrae so you can recognize them in the disarticulated skeleton.

Now examine the appendicular skeleton. In comparison to a lower vertebrate such as a frog, the cat's limbs are positioned more under the body. This occurred through the rotation of the elbow of the forelimb posteriorly and rotation of the knee of the hind limb anteriorly. Spend some time studying the articulations of the forelimb and hind limb bones. Notice that the cat walks on its toes (digits). This is called a *digitigrade* stance. The contact of the foot with the ground tends to be consistent and distinctive in each mammalian order. For example, ungulates, such as deer, walk on hoofs (modified claws), and primates walk on the entire sole of the foot.

For the rest of the skeletal study, use the disarticulated skeleton. Refer to Figure 2 and the mounted skeleton as needed to determine the correct articulation and positioning of the bones.

The Skull

Refer to Figures 3-5 for the bones of the skull. Most of these bones articulate by immobile joints or *sutures*. Functionally, the two largest divisions of the skull are the *cranium,* which encloses the brain, and the facial bones, which support the eyes, nose, and

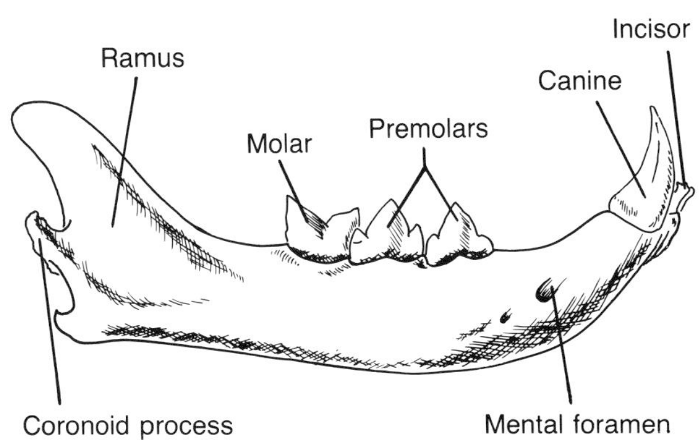

Figure 5. Mandible (right lateral view).

mouth. Trace the limits of each bone with a dissecting needle.

Look into the external nares. There you will see the many thin, folded plates of the *turbinate bones*. In life, these support and are covered by the *olfactory epithelium*. The turbinates are located in the nasal passages, which open posteriorly as the *internal nares*. Now look through the *foramen magnum* (*foramen:* an opening) into the cranial cavity. The spinal cord passes through this foramen. If you have a sectioned skull you will be able to see that the cranial cavity is incompletely divided into three chambers. Beginning at the posterior of the skull, these are the *cerebellar fossa* (*fossa:* a pit or depression), which houses the cerebellum, the *cerebral fossa,* which houses the cerebrum, and the *olfactory fossa,* a small chamber at the extreme anterior end. The olfactory fossa houses the olfactory bulbs. Notice that the anterior wall of the olfactory fossa is a bony plate penetrated by many small foramini. This is the *cribriform plate*. Bundles of *olfactory nerves* pass through the cribriform plate to connect the olfactory bulbs of the brain with sensory nerve endings in the olfactory epithelium. You will notice many other foramini in the skull. Most of these are passageways for nerves or blood vessels.

Returning to the exterior of the skull, examine the *orbit* or eye socket. The small foramen at the anterior edge of the orbit is the *lacrimal canal* (nasolacrimal duct or tear duct). The larger *infraorbital foramen* is just ventral to the lacrimal canal opening. Below and lateral to the orbit is the *zygomatic arch,* which consists of parts of the zygomatic and temporal bones. Find, on the ventral posterior portion of the zygomatic arch, the *mandibular fossa* where the mandible articulates. The *temporal fossa* is the large opening posterior to the orbit and incompletely separated from it by the *postorbital* and *frontal processes* (*process:* a projection).

The *tympanic bulla* contains the *malleus, incus,* and *stapes,* the bones of the middle ear. The *external auditory meatus* is the oval opening near the posterior base of the zygomatic arch. Unless the malleus has been removed, it should be visible within the tympanic bulla.

Now examine the teeth. It is often possible to identify a mammalian species by the teeth alone, and the dental pattern tells us much about the mammal's diet. Beginning at the anterior end of the mouth, the cat has *incisors, canines, pre-*

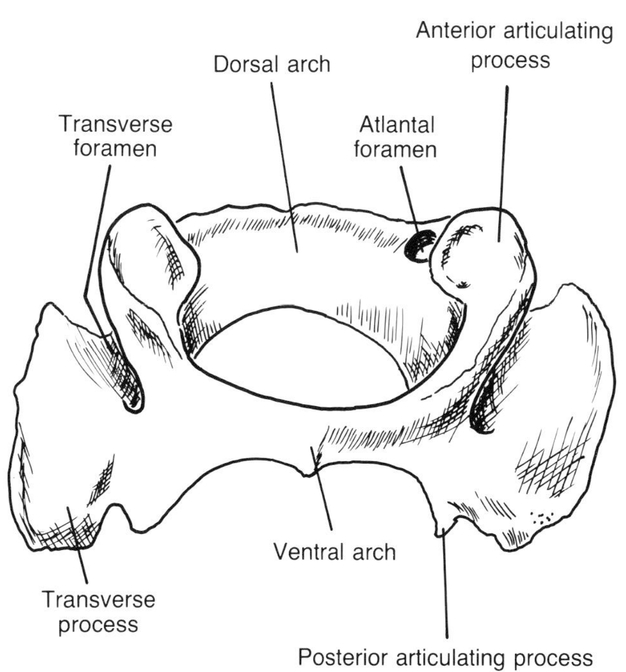

Figure 6. Atlas (ventral view).

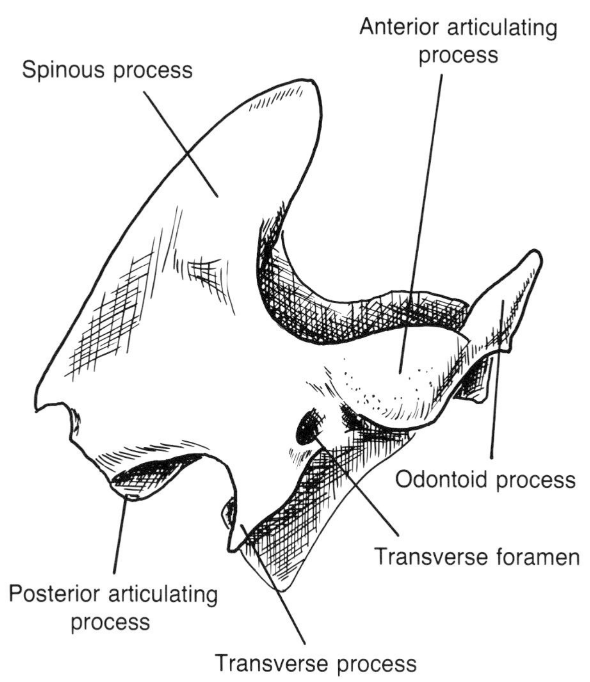

Figure 7. Axis (lateral view).

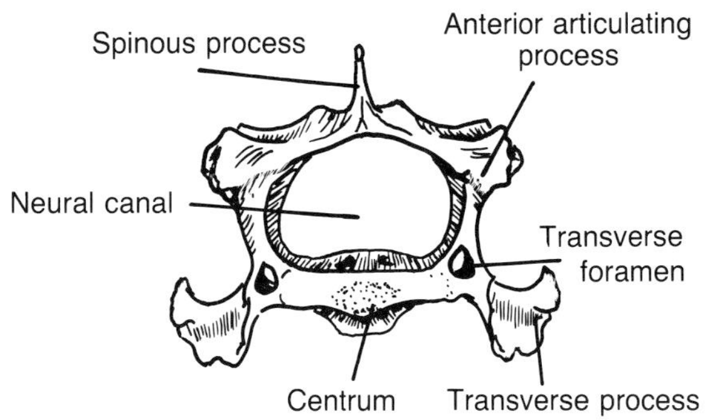

Figure 8. Cervical vertebra (anterior view).

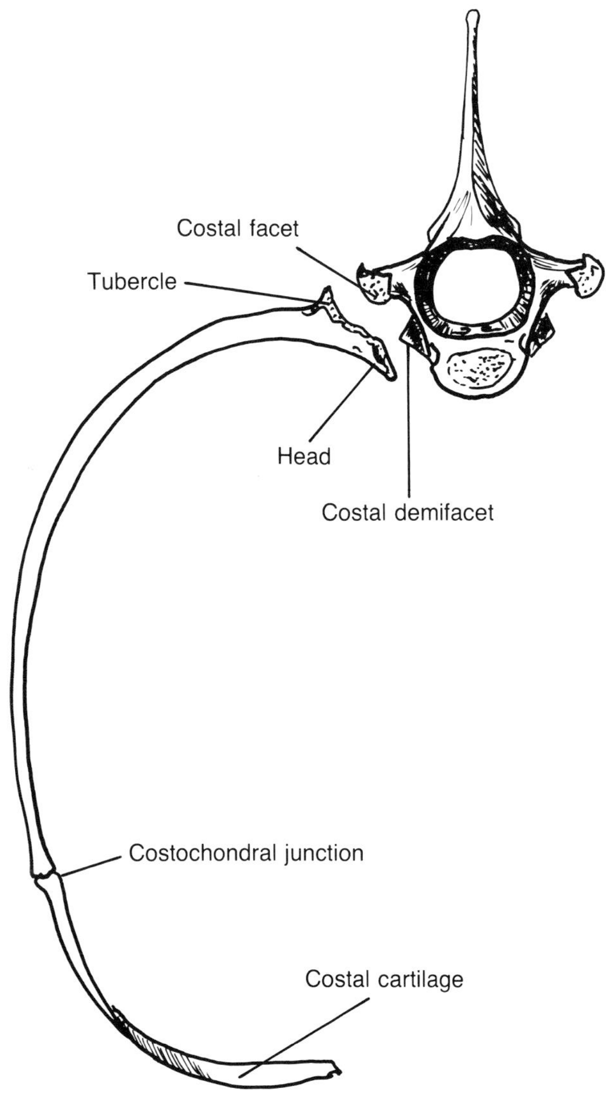

Figure 9. Thoracic vertebra and rib (posterior view).

molars, and *molars.* The dental formula for an adult human is I $\frac{2}{2}$, C $\frac{1}{1}$, P $\frac{2}{2}$, M $\frac{3}{3}$. This indicates that an adult human has two incisors on each side of the upper jaw and two incisors on each side of the lower jaw, etc. Determine the dental formula for your specimen. Most mammals have two sets of teeth, deciduous ("baby" or "milk" teeth) and permanent teeth.

The Vertebrae and Rib Cage

Refer to Figures 6-10. The vertebrae, except for the atlas and the caudals, have the following features in common: a *centrum,* lateral *transverse processes,* a dorsal *spinous process,* a *neural canal,* and *articular processes.* In the case of the atlas, its centrum has fused to the axis to form the *odontoid process* (Fig. 7), and there is no spinous process. The vertebral artery and vein pass through the *transverse foramen* of the atlas and follow a groove to the *atlantal foramen.* This latter foramen is also a passageway for the first spinal nerve. If you re-examine the mounted skeleton, you will find that there is a space or *intervertebral foramen* between the members of each pair of articulating vertebrae for the passage of the spinal nerves.

Each thoracic vertebra articulates with one or two pairs of ribs. One articulation is to a facet on the transverse process (*facet:* a small, smooth articulating surface). The second is to a facet on the centrum. This latter facet usually consists of two parts, one being on the posterior edge of a centrum and the other being on the anterior edge of the following centrum. In those cases it is called a *demifacet.*

The distal portion of each rib is of cartilage. The cartilaginous portions of the first nine pairs of ribs, called *true ribs,* articulate with the *sternum* or breastbone. The four posterior ribs that do not articulate with the sternum are called *false ribs.* The first three of these false ribs articulate with other ribs, but the fourth has no distal articulation and is called a *floating rib.*

The sternum, like the backbone, is segmented. The segments are called *sternebrae.* The most anterior sternebra is further distinguished as the *manubrium.* The most posterior sternebra is the *xiphisternum,* which consists of bone and the *xiphoid cartilage.*

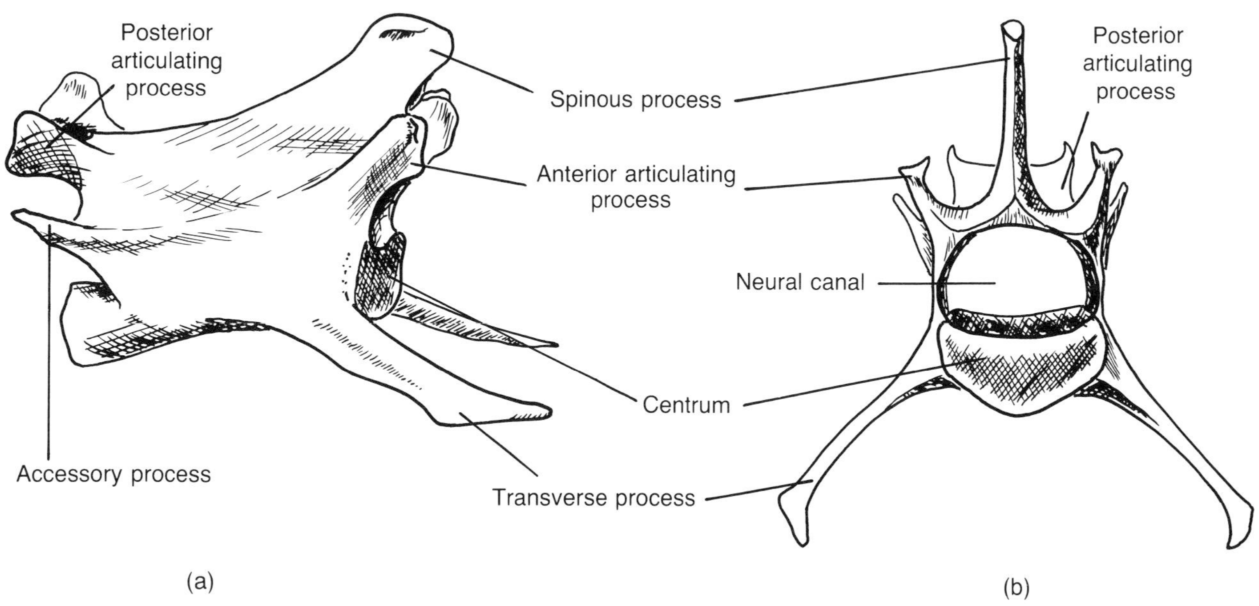

Figure 10. Lumbar vertebra: (a) right lateral view; (b) anterior view.

The lumbar vertebrae show an increase in length of the transverse process, and articulate to give good dorsal-ventral flexibility. The sacrum shares most of its features with the other vertebrae. Other than the fusion of its parts, perhaps its most distinguishing feature is the enlargement of its articular surfaces. The caudal vertebrae become much simplified as one proceeds posteriorly. The *neural canal* (the tube formed by the vertebral canals) disappears and the processes are reduced to knobs. What remains is the centrum. The *chevron bones* are interesting features of the caudal vertebrae. These small bones may be found attached to the ventral surface of the sixth, seventh, and eighth caudal vertebrae where they are fused in a "v" shape. They form an arch that may be a remnant of the *hemal arch* of more primitive vertebrates such as the dogfish.

The Pectoral Girdle and Forelimb

The pectoral girdle consists of the *scapulae* (Fig. 11) and *clavicles.* In man, the clavicles or collarbones form a sturdy brace for the shoulders, but in the cat, and in the Carnivora in general, the clavicles are reduced and articulate with no other bone. Notice that the scapula does not articulate with the backbone. Instead, the scapula is held in place by strong muscles. The flat medial (inner) surface of the

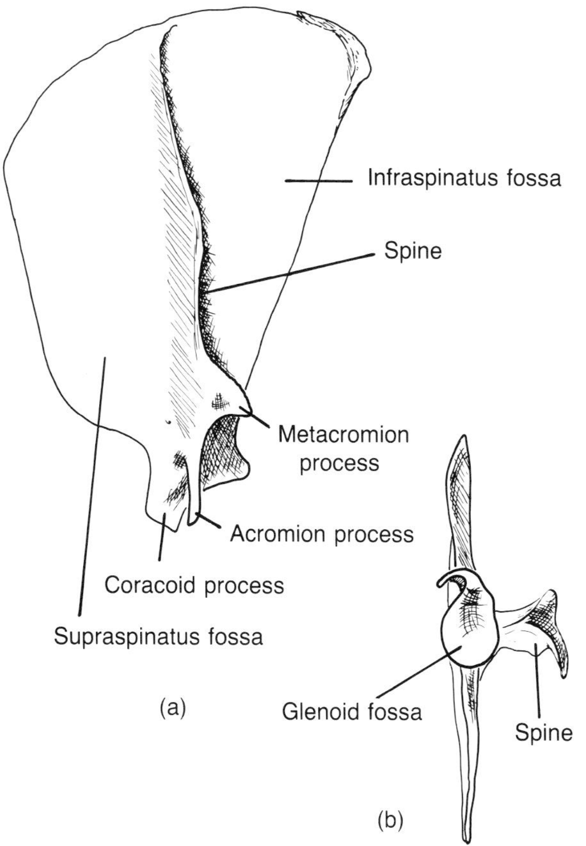

Figure 11. Left scapula: (a) lateral view; (b) anterior view.

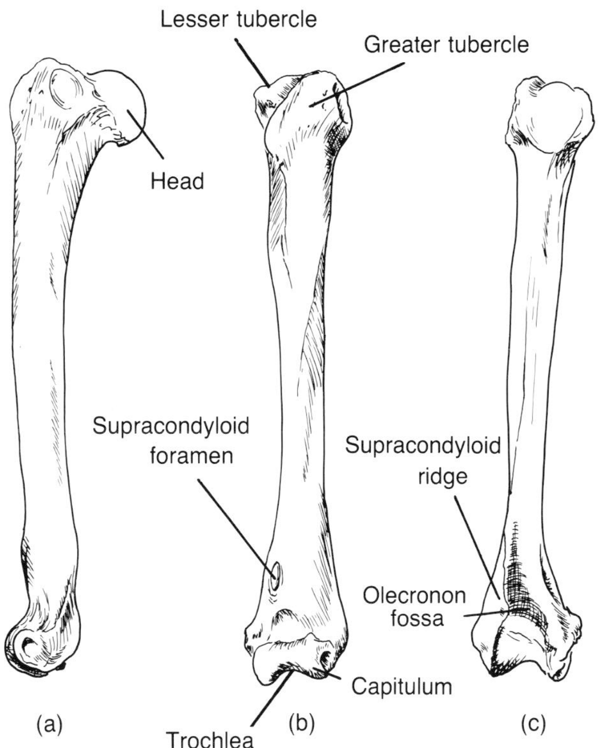

Figure 12. Left humerus: (a) lateral view; (b) anterior view; (c) posterior view.

scapula is the *subscapular fossa*. The *coracoid process* is homologous to the coracoid bone found in many lower vertebrates.

The proximal end of the *humerus* (Fig. 12) has two processes or tubercles. A muscle tendon slides through the *intertubercular groove* between them. Near the distal end is the *supracondyloid foramen,* a passageway for the brachial artery and median nerve.

The *ulna* is the longest bone of the forelimb. The *radius* and ulna (Fig. 13) form a broad surface for articulation with the *carpals*. The *metacarpals* and *phalanges* are numbered one through five beginning with the *pollex* (thumb). We will consider the bones of the foot as one complete articulated unit. Figure 14 shows how the claws are retracted or extended.

Practice articulating the bones of the forelimb (except the individual bones of the foot) until you are able to do so easily without referring to any notes.

The Pelvic Girdle and Hind Limb

The pelvic girdle consists of two bones, the *innominates,* which are fused ventrally along the *pu-*

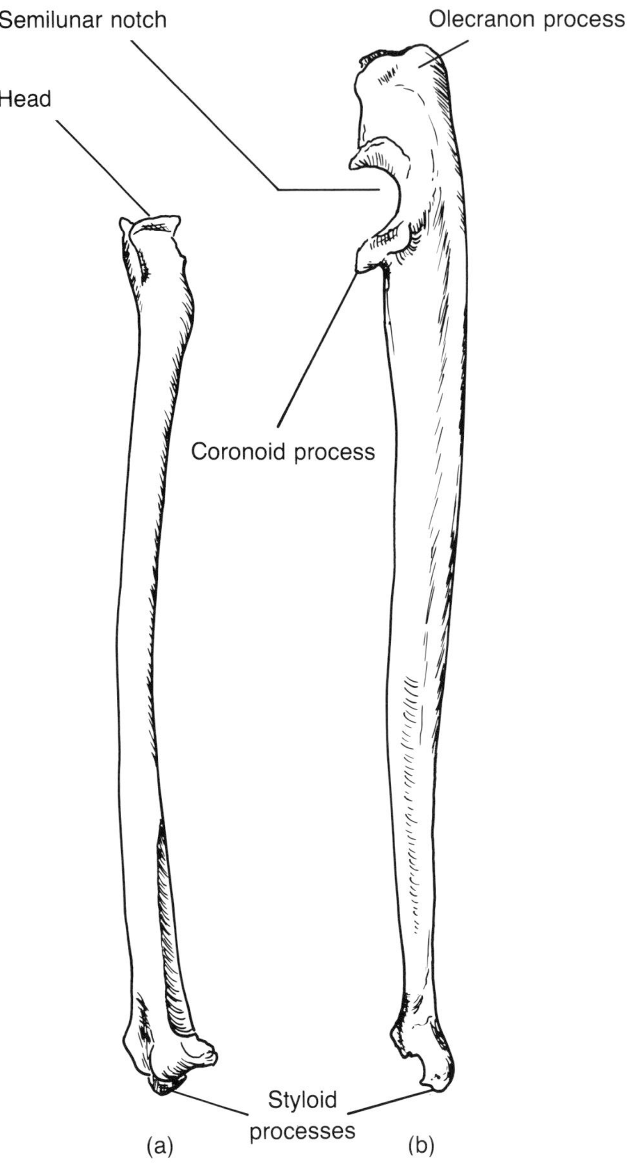

Figure 13. (a) Left radius (lateral view); (b) left ulna (lateral view).

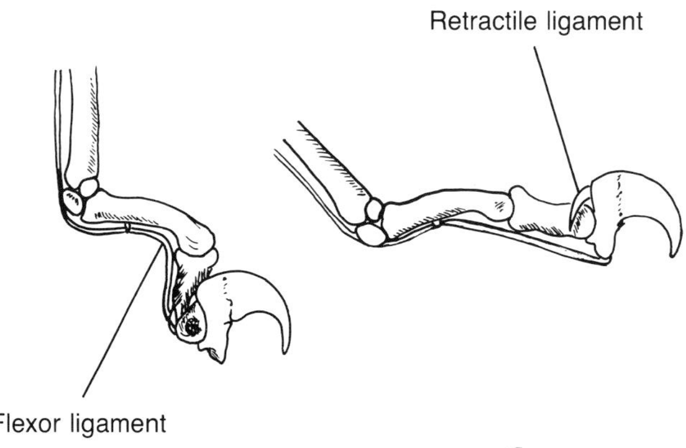

Figure 14. Extension of the claw.

14

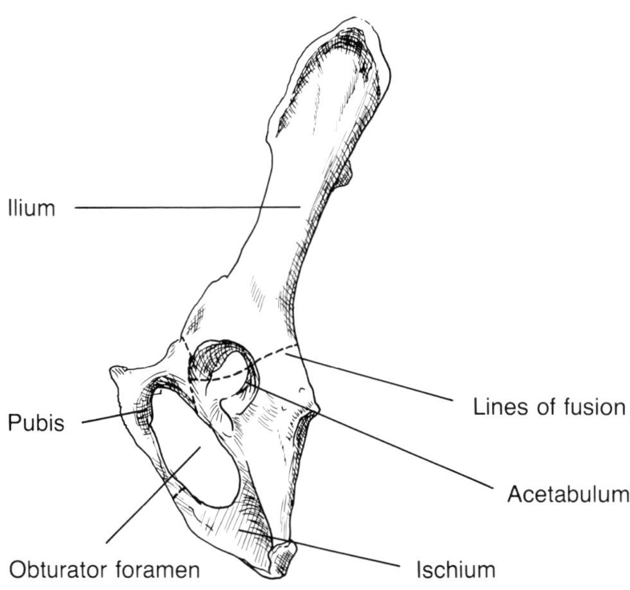

Figure 15. Left innominate (lateral view).

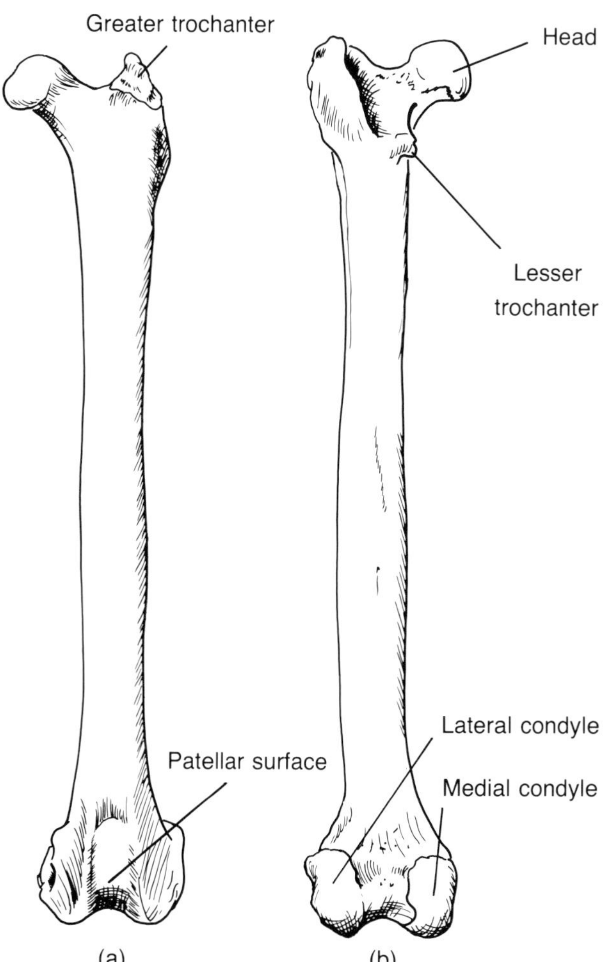

Figure 16. Left femur: (a) anterior view; (b) posterior view.

bic symphysis and articulate medially with the spinal sacrum. Each innominate is formed by three fused bones. The limits of each bone are shown by dotted lines on Figure 15, but your specimen will show no traces of fusion. The *obturator foramen* allows the passage of nerves and blood vessels, and the *acetabulum* accepts the head of the *femur*. Examine the head of the femur (Fig. 16) and you will find a pit that was the attachment site for a ligament that held the head in the acetabulum.

The *patella* is the largest *sesamoid bone* in the body. (A sesamoid bone is one that develops in a tendon.) The *tibia* (Fig. 17) is the body's longest bone. It articulates distally with a surface on the *talus,* a characteristically shaped tarsal bone. The heel, which projects posterior to the ankle joint, is another tarsal bone, the *calcaneus*. The *fibula* is positioned laterally to the tibia.

Practice articulating the bones of the hind limb (except the individual bones of the foot) until you are able to do so easily without looking at any reference.

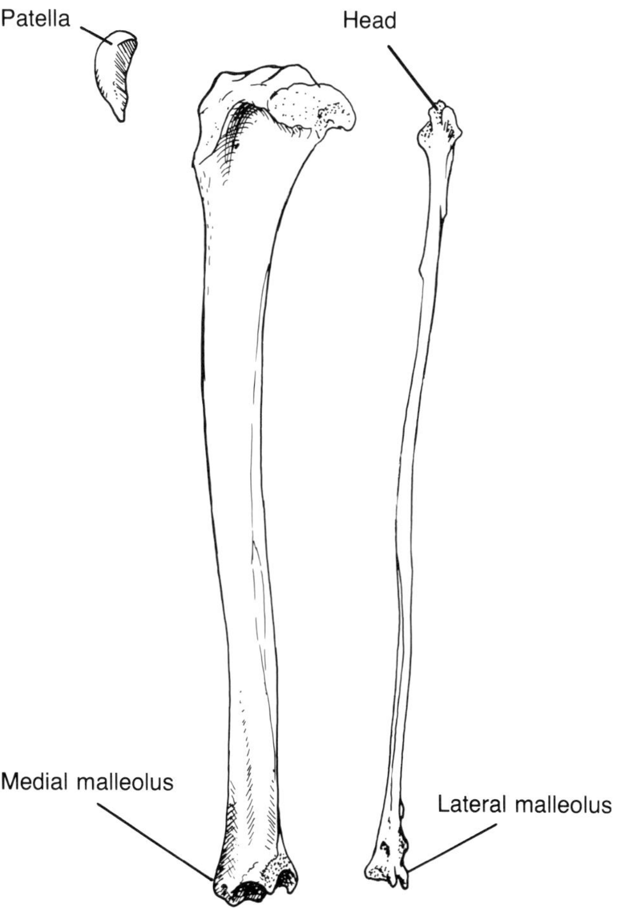

Figure 17. Left tibia and fibula (lateral view).

Questions and Activities

1. What is the dental formula for an adult cat?
2. Describe the movement allowed by the axis-to-atlas joint and the atlas-to-skull joint.
3. What is the usual number of mammalian cervical vertebrae? Give some exceptions.
4. Give the articulating surfaces for the following forelimb joints: shoulder, elbow, wrist.
5. Give the articulating surfaces for the following hind limb joints: hip, knee, ankle.

Unit 3
THE MUSCLES

Skinning

Use at least one lab period to skin the cat. You need forceps, scissors, and a blunt probe (not a needle). The blunt end of a scalpel handle can be used as a probe if you first remove the blade. Otherwise, do not use the scalpel for muscle dissection as it can damage your specimen.

Place the cat on its back. There should be a cut on the neck where blood vessels were injected. Beginning at that point, lift the skin with forceps, insert the blunt tip of your scissors and cut forward to the "chin" (see Fig. 18). Notice that a white, fibrous tissue, the *superficial fascia,* connects the skin to the underlying tan-colored muscles. Now extend the cut posteriorly along the ventral midline to the external genitalia. Cut around the genitalia, the anus, and the base of the tail. Extend the cut down each limb and around each "wrist" just before the paw. Also cut around the mouth to the eyes and around the eyes and ears. As you cut the skin, lift it up and use the blunt probe to separate the skin from the muscle. In some places you may have to use the scissors to get the skin to separate properly. Work with the tip of the probe or scissors against the skin rather than against the muscles. This will prevent damage to the muscles. Remove the skin as a whole piece rather than cutting it into bits. Use the scissors to cut the mammary glands from the skin as close to the nipples as possible.

As you skin the thorax, you will find the *cutaneous maximus,* the cutaneous muscle that twitches the skin. In the neck area is a second cutaneous muscle, the *platysma.* It is almost impossible to remove the skin without damaging these muscles, and they must be removed before the skeletal muscles can be studied; however, because of the danger of tearing a deeper muscle, try to leave the cutaneous muscles intact for now.

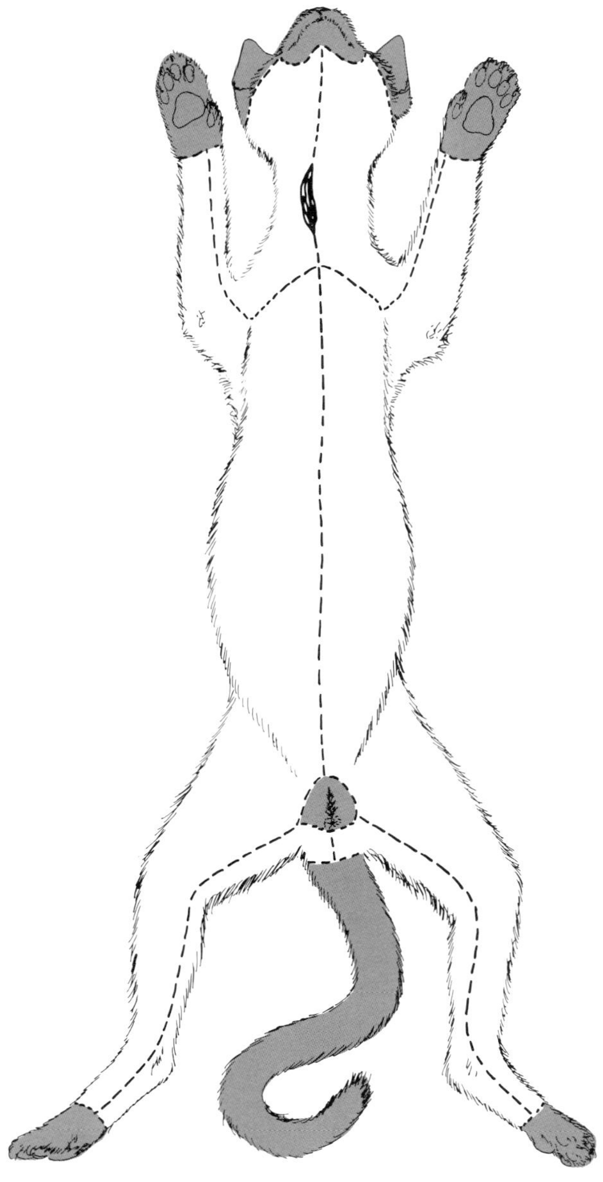

Figure 18. Skinning diagram.

Table 1
Chest Muscles

Muscle	Origin	Insertion	Action
Biceps brachii	Scapula near glenoid fossa	Radial tuberosity of radius	Flexes forearm
Coracobrachialis	Coracoid process	Medial surface of humerus	Adducts arm
Clavobrachialis	Clavicle	Ulna	Flexes forearm
Clavotrapezius	Dorsal midline of neck	Clavicle	Flexes forearm
Levator scapulae	Cervical vertebrae	Scapula	Pulls scapula anteriorly
Levator scapulae ventralis	Posterior of skull	Scapula	Pulls scapula anteriorly
Longus capitis	2nd to 6th cervical vertebrae	Occipital bone of skull	Depresses snout
Occipitoscapularis	Posterior of skull	Scapula	Pulls scapula anteriorly
Pectoantibrachialis	Manubrium	Elbow, superficial	Adducts arm
Pectoralis major	Midventral line	Humerus	Adducts arm
Pectoralis minor	Sternum	Proximal end of humerus	Adducts arm
Rectus abdominis	1st and 2nd ribs	Pubis	Compresses abdomen
Scalenus		Cervical vertebrae	Flexes neck
Scalenus dorsalis	3rd or 4th rib		
Scalenus medius	6th to 9th ribs		
Scalenus ventralis	2nd and 3rd ribs		
Serratus ventralis	Ribs 1 through 10	Vertebral border of scapula	Depresses scapula
Splenius	Between spines of cervical and first two thoracic vertebrae	Occipital	Turns or elevates head
Subscapularis	Subscapular fossa	Lesser tuberosity of humerus	Adducts arm
Teres major	Posterior border of scapula	Humerus	Flexes humerus
Transverse costarum	Middle sections of sternum	1st rib	Pulls sternum forward
Xiphihumeralis	Xiphisternum	Proximal end of humerus	Adducts arm

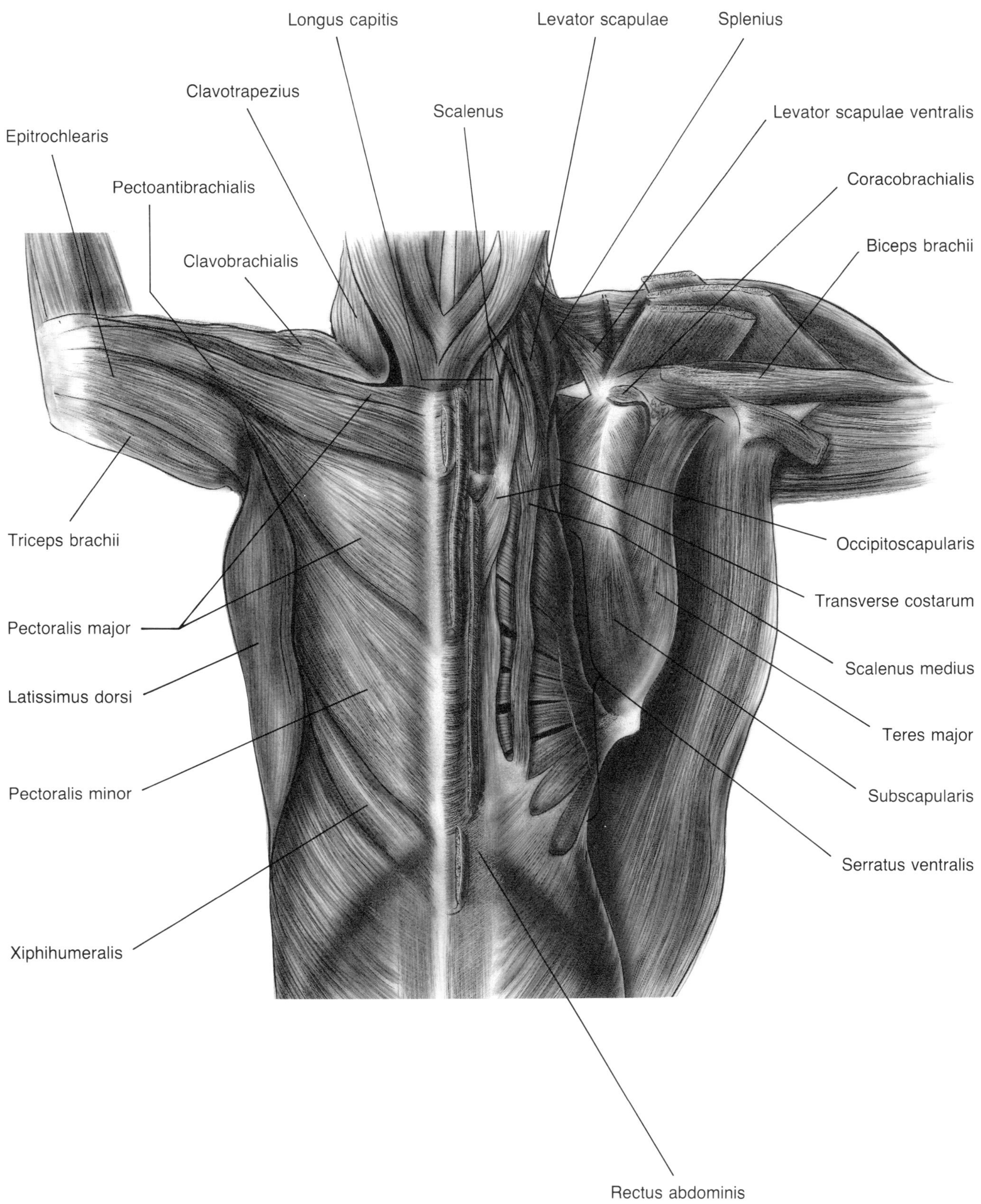

Figure 19. Chest muscles, superficial and deep (ventral view).

General Structure and Action

A muscle consists of a fleshy portion or belly, which is attached at both ends to a bone or another muscle by connective tissue. The connective tissue may be a *tendon* or a flattened band called an *aponeurosis*. When a muscle contracts, one end usually remains more-or-less fixed while the other end moves. The fixed end is called the *origin* and the moveable end is the *insertion*. Muscle origins are usually medial or proximal. Insertions are usually lateral or distal.

Muscles can be classified according to their actions as follows.

Flexor — bends a joint.

Extensor — straightens a joint.

Abductor — moves a part away from the median plane.

Adductor — moves a part toward the median plane.

Levator — raises a part.

Depressor — lowers a part.

Rotator — twists or rotates a part. *Pronators* turn the dorsal surface of a limb to the anterior, and *supinators* turn the ventral surface of a limb to the anterior.

Sphincter — closes or constricts an orifice.

Dilator — opens an orifice.

As you might guess from the above list, muscles tend to work in opposing pairs. This is because a muscle works only when it contracts; it cannot exert force when it relaxes.

Muscle Dissection

At this point you may wonder, "Where are the muscles?" Certainly there are few signs of the clear, distinct, rounded muscles shown on charts. Instead, the muscles are covered and held tightly together by connective tissue. Therefore, careful, slow work is required to expose them.

Use forceps, blunt probe, and scissors to remove the fat and connective tissue covering the muscles and to separate the belly of each muscle from its neighbors. Dissect the muscle groups in the order in which they are shown in this manual. The drawings are to help you locate and identify the muscles. You will see more detail than we can present here. In a few cases, an aponeurosis or other tissue has been removed to show underlying structures.

It may be difficult to find the separation between two adjacent muscles. If so, use caution or you may "create" a muscle. It is often better to begin at the origin or insertion and work toward the belly.

It will be necessary to transect and reflect (fold back) superficial muscles to expose deeper muscles. This is accomplished by cutting through the belly of the superficial muscle and reflecting its cut halves towards the muscle's origin and insertion respectively. The superficial muscles can then be replaced for review or further study. (To show the relative positions of superficial and deeper muscles, portions of superficial muscles have been completely removed for some of the illustrations.) We recommend that deep muscle dissection be done on the left side only. Then, in case of a major error, the right side will be available. This will also leave the right blood vessels and nerves intact for later dissection.

Once the muscles are well exposed, trace their origins and insertions. Origins, insertions, and actions for each of the muscles you will dissect are listed on the following pages. These are somewhat generalized. The exact surface features of the bone on which a muscle has its origin are not listed. Refer to the articulated skeleton or Figure 2 when studying origins and insertions. The muscle actions listed are based only on the contraction of the single muscle under consideration. Actually, a muscle usually works with other muscles to produce other actions in addition to those listed. It will be helpful to think of how groups of muscles move a joint or limb.

Chest muscles *(Table 1 and Figure 19).* The chest muscles are large and easy to dissect, making the chest a good place to start. In exposing the deep chest muscles, you will find nerves and blood vessels coming out of the lateral chest wall and going into the arm. These have been removed to produce the drawing of the deep chest muscles, but you should leave them intact if possible. Try to leave all blood vessels and nerves intact unless specifically told to remove them.

Table 2
Head and Neck Muscles

Muscle	Origin	Insertion	Action
Cricothyroid	Cricoid cartilage	Thyroid cartilage	Retracts thyroid cartilage
Digastric	Base of skull	Mandible	Depresses mandible
Geniohyoid	Mandible	Hyoid	Pulls hyoid anteriorly
Hyoglossus	Hyoid	Tongue	Retracts tongue
Masseter	Zygomatic arch	Mandible	Elevates mandible
Mylohyoid	Mandible	Midventral raphe	Elevates floor of mouth
Sternohyoid	Manubrium	Hyoid	Pulls hyoid posteriorly
Sternomastoid (caudal and cranial)	Manubrium	Posterior of skull	Turn head and depress snout
Sternothyroid	1st rib	Thyroid cartilage	Pulls larynx posteriorly
Stylohyoid	Wing of hyoid	Body of hyoid	Elevates hyoid
Thyrohyoid	Thyroid cartilage	Hyoid	Pulls hyoid posteriorly

Transect and reflect the following to expose the deep chest muscles: pectoantibrachialis, pectoralis major superficial and deep (the superficial is covered by the pectoantibrachialis, to which it is similar in size and orientation), pectoralis minor, xiphihumeralis, clavobrachialis.

The levator scapulae ventralis, pectoantibrachialis, transverse costarum, and xiphihumeralis have no human homologues. The clavobrachialis, together with the acromiodeltoid and spinodeltoid (to be studied later), is homologous to the human deltoid. Likewise, the clavotrapezius, acromiotrapezius, and spinotrapezius (the latter two studied later) are collectively homologous to the single human trapezius. The pectoralis major may seem misnamed in the cat, but in humans the pectoralis major is much larger than the pectoralis minor.

Head and neck muscles *(Table 2 and Figure 20).* Also shown here are three of the *salivary glands* and the *thyroid gland*. Although there are other salivary glands, they are small in the cat and difficult to find. The salivary glands are easily confused with the large *lymph nodes* in the area; however, the lymph nodes are smooth while the salivary glands are bumpy. The thyroid gland is part of the endocrine system. You will see the thyroid better in later dissections. This area of the neck was cut for injection of the blood vessels, and the resulting damage determines the exact procedure used to expose the deep muscles. Generally the sternomastoid (cranial), stylohyoid, and sternohyoid must be transected and reflected. In Figure 20 the mylohyoid is pushed anteriorly and the thyrohyoids and sternothyroids are pushed laterally.

21

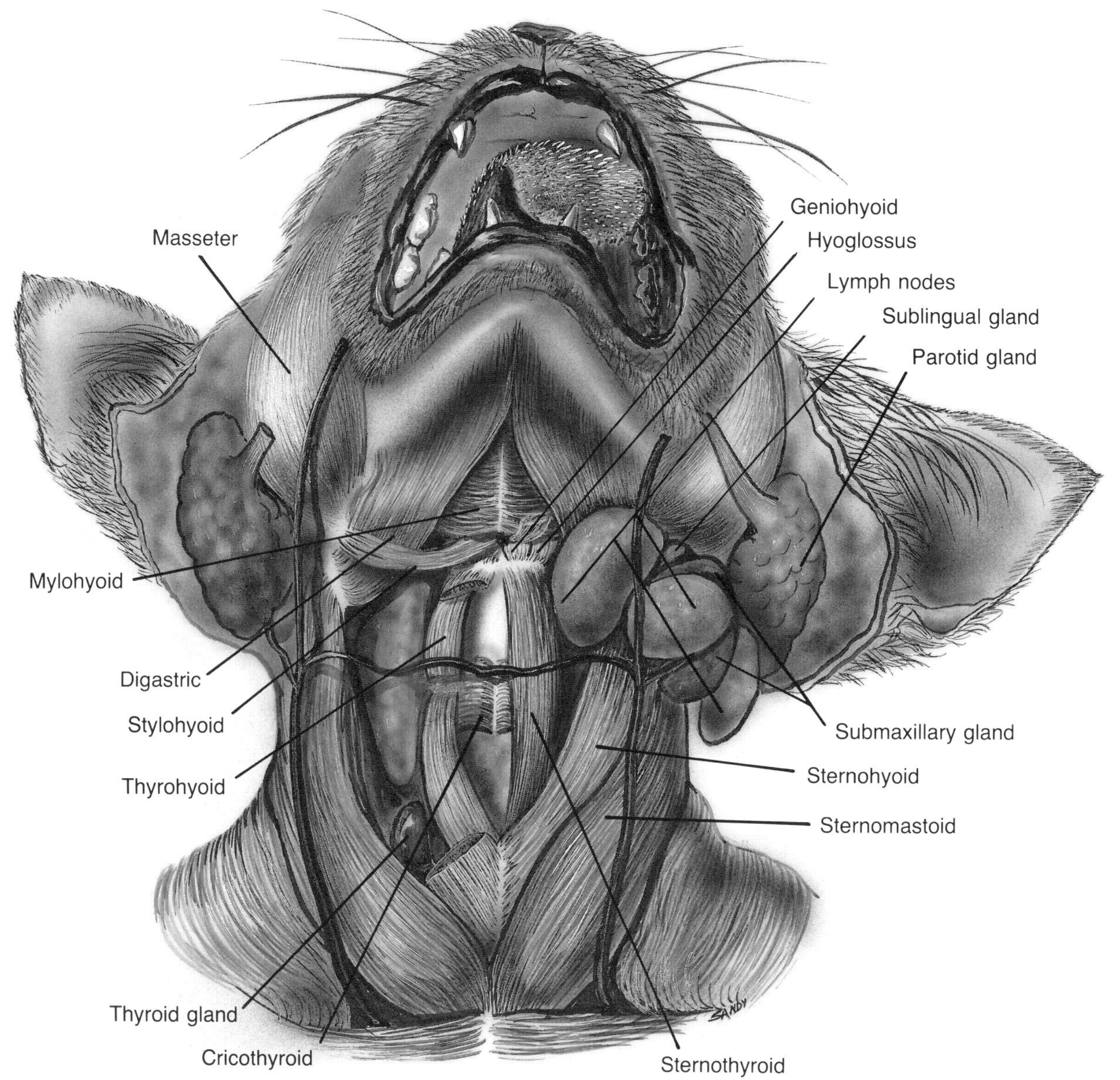

Figure 20. Head and neck muscles (including glands of the throat).

Forelimb muscles *(Table 3)*. The dissection of the forelimb muscles requires extra care because they are small and tightly packed. A tough sleeve of connective tissue tightly covers the forearm muscles and must be removed. Become fully familiar with the superficial muscles (Figs. 21 and 22) before trying the deep muscles.

Transect and reflect the following to expose the deep muscles of the medial forelimb (Fig. 23): triceps (lateral head), brachioradialis, extensor digi-torum communis, extensor digitorum lateralis, flexor carpi ulnaris.

The following must be transected and reflected for the deep muscles of the lateral forelimb (Fig. 24) to be seen: epitrochlearis, palmaris longus, flexor carpi ulnaris, flexor carpi radialis.

The anconeus and epitrochlearis have no human homologues. The extensor carpi radialis brevis, extensor digitorum lateralis, and palmaris longus are relatively smaller in humans, and the palmaris long-us is missing in about 10% of the human population.

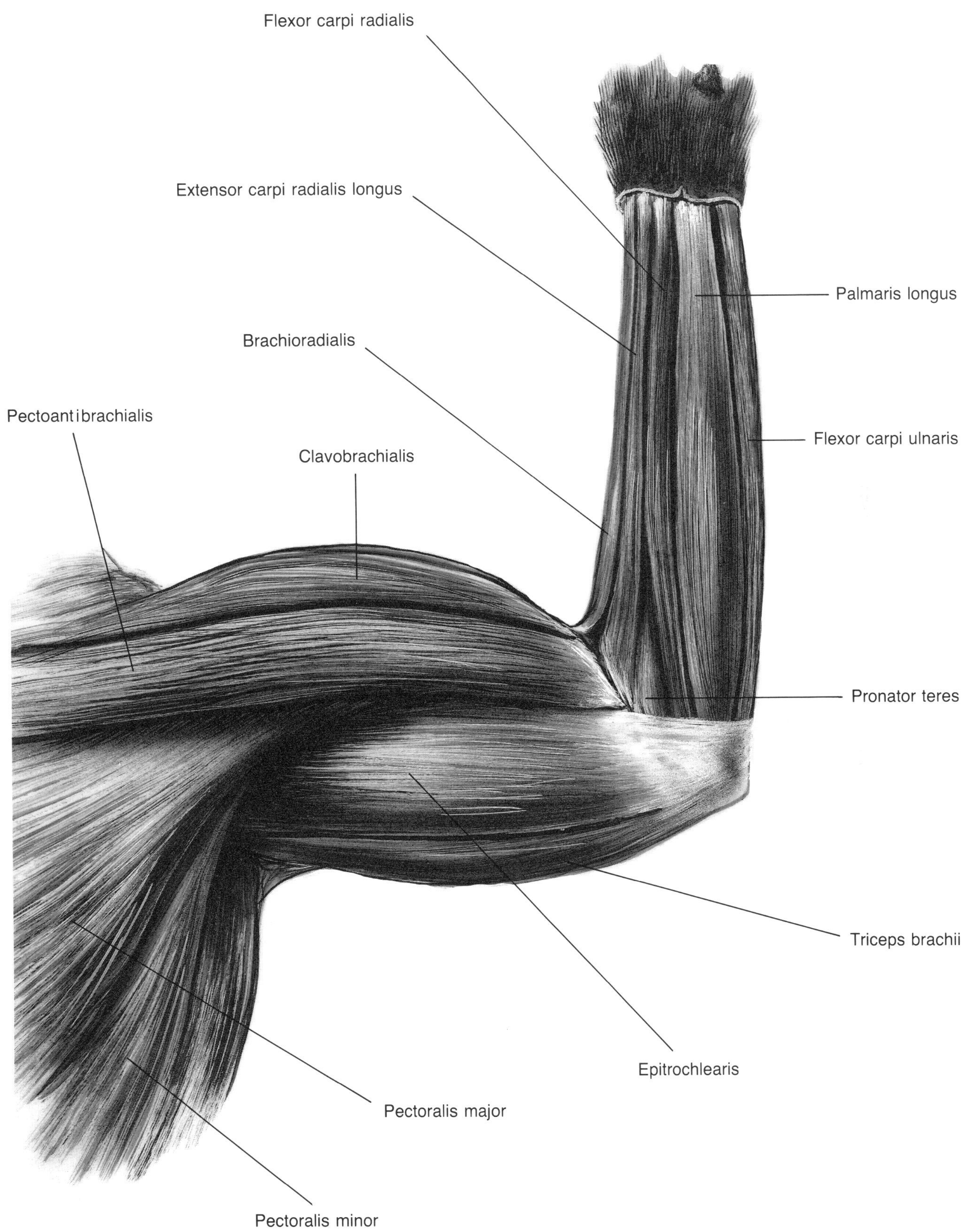

Figure 21. Left forelimb muscles, superficial (medial view).

23

Table 3
Forelimb Muscles

Muscle	Origin	Insertion	Action
Acromiodeltoid	Acromion of scapula	Humerus	Flexes humerus
Abductor pollicus longus	Ulna and radius	1st metacarpal	Abducts 1st digit
Anconeus	Distal end of humerus	Lateral, proximal end of ulna	Tenses elbow joint
Brachialis	Lateral surface of humerus	Proximal end of ulna	Flexes forearm
Brachioradialis	Dorsal surface of humerus	Styloid process of radius	Supinates hand
Epitrochlearis	Ventral border of latissimus dorsi	Olecranon process of ulna	Extends forearm
Extensor carpi radialis brevis	Distal end of humerus	3rd metacarpal	Extends hand
Extensor carpi radialis longus	Distal end of humerus	2nd metacarpal	Extends hand
Extensor carpi ulnaris	Distal end of humerus	5th metacarpal	Extends wrist
Extensor digitorum communis	Distal end of humerus	2nd and 3rd phalanges	Extends digits
Extensor digitorum lateralis	Distal end of humerus	3rd to 5th phalanges	Extends digits
Extensor pollicus longus	Lateral surface of ulna	1st phalanx	Extends 1st digit
Flexor carpi radialis	Distal end of humerus	2nd and 3rd metacarpals	Flexes wrist
Flexor carpi ulnaris	Distal end of humerus and olecranon process of ulna	A carpal bone	Flexes wrist
Flexor digitorum profundus	Several heads with origins mainly on the ulna and humerus	Phalanges	Flexes digits
Palmaris longus	Distal part of humerus	1st to 5th phalanges	Flexes digits
Pronator teres	Distal end of humerus	Medial border of radius	Pronates hand
Supinator	Ligaments of lateral elbow joint	Proximal end of radius	Suppinates hand
Triceps brachii		Olecranon process of ulna	Extends forearm
Lateral head	Proximal end of humerus		
Long head	Border of glenoid fossa		
Medial head (intermediate and short heads)	Medial surface of humerus		

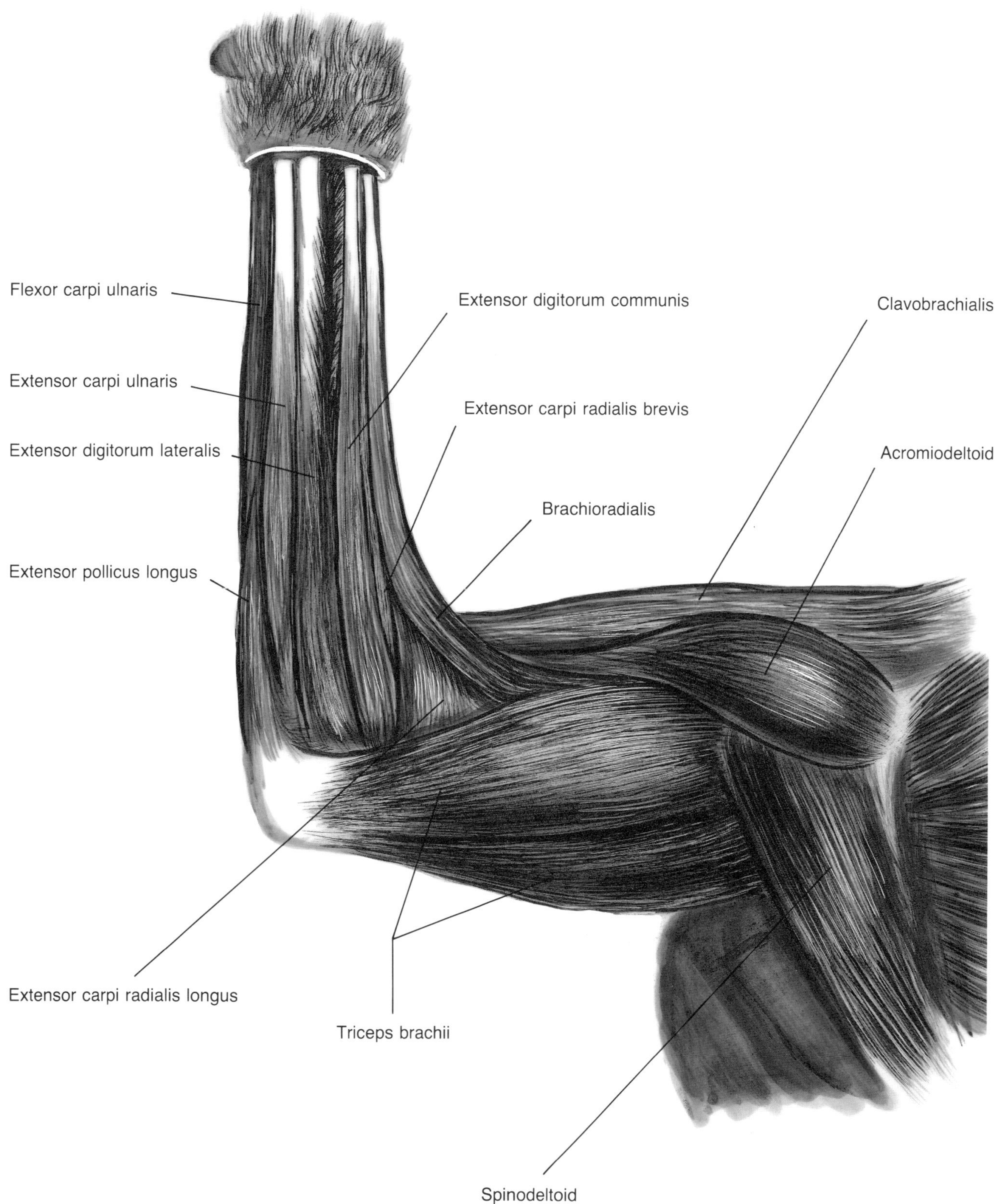

Figure 22. Left forelimb muscles, superficial (lateral view).

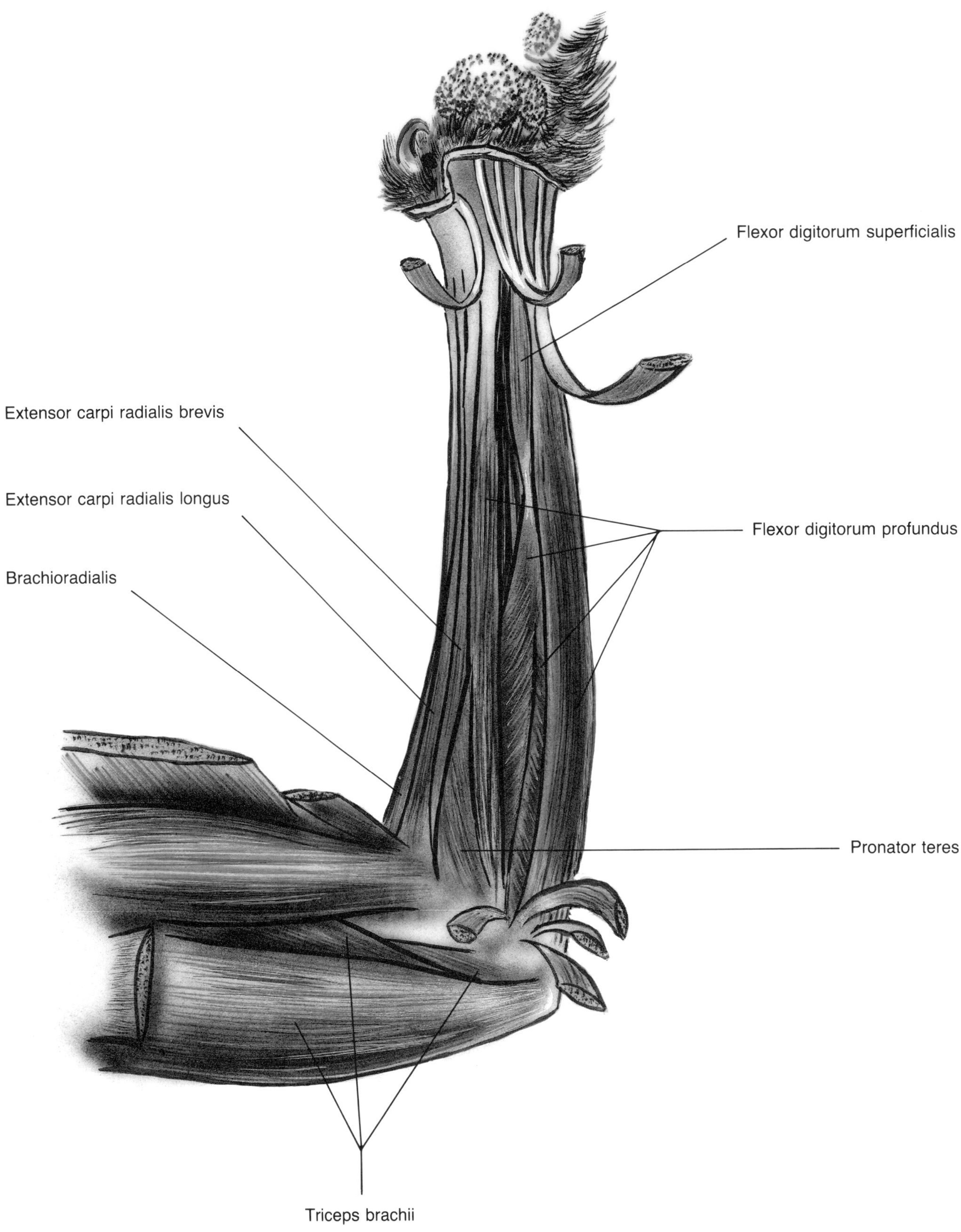

Figure 23. Left forelimb muscles, deep (medial view).

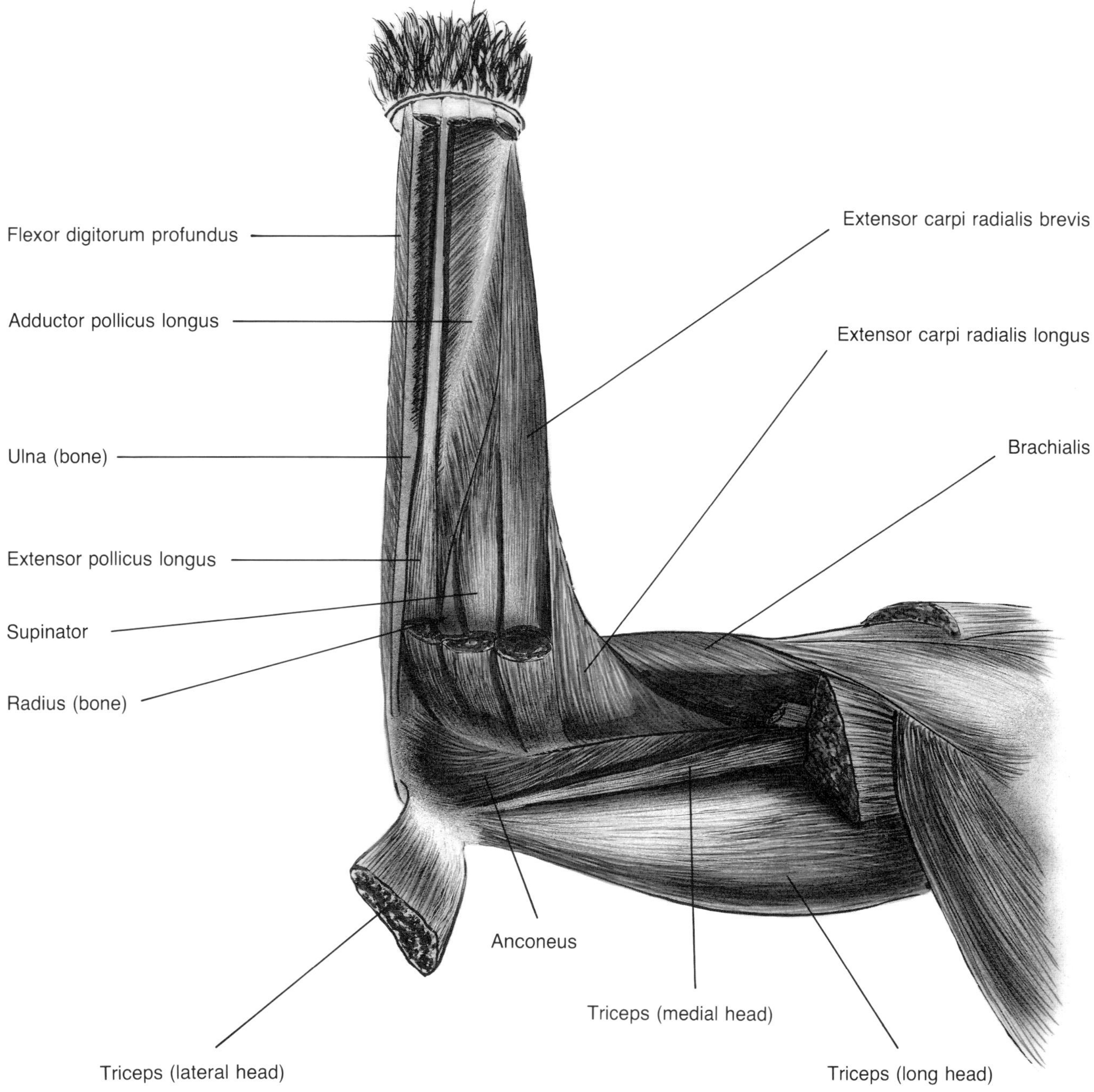

Figure 24. Left forelimb muscles, deep (lateral view).

Hind limb muscles *(Table 4)*. Dissecting the hind limb muscles presents few problems other than identifying some muscles of the lower leg. Once the superficial muscles (Figs. 25 and 26) are known, transect and reflect the biceps femoris to expose the deep muscles of the lateral hind limb (Fig. 27). The delicate tenuissimus will be torn by this procedure.

Do not try to trace the origin or insertion of the tenuissimus as this is difficult. Transect and reflect the sartorius and gracilis to expose the deep medial hind limb muscles (Fig. 28).

The caudofemoralis and tenuissimus are without human homologues, and the gluteus maximus is relatively larger in humans.

Table 4
Hind Limb Muscles

Muscle	Origin	Insertion	Action
Adductor femoris	Pubis	Femur	Adducts thigh
Adductor longus	Pubis	Femur	Adducts thigh
Biceps femoris	Ischium	Tibia and patella	Extends thigh
Caudofemoralis	2nd and 3rd caudal vertebrae	Patella	Abducts thigh
Extensor digitorum longus	Distal end of femur	2nd to 5th phalanges	Extends digits 2-5
Flexor digitorum longus	Tibia	2nd to 5th phalanges	Flexes digits
Gastrocnemius	Condyles of femur	Calcaneus	Extends foot
Gemellus inferior	Ischium	Tendon of deeper muscle	Abducts thigh
Gluteus maximus	Fibers and fascia of the posterior spinal area	Greater trochanter	Abducts thigh
Gluteus medius	Ilium and fascia	Greater trochanter	Abducts thigh
Gracilis	Pubis	Proximal end of tibia	Abducts thigh
Iliopsoas	Lumbar vertebrae, ilium	Proximal end of femur	Flexes thigh
Pectineus	Pubis	Proximal end of femur	Adducts thigh
Peroneus longus	Proximal half of fibula	Metatarsals	Flexes foot
Peroneus tertius	Lateral surface of fibula	5th phalanx	Abducts 5th digit
Quadratus femoris	Ischium	Greater trochanter	Extends thigh
Rectus femoris	Ilium	Patella	Extends lower leg
Sartorius	Ilium	Proximal end of tibia	Flexes thigh
Semimembranosus	Ischium	Distal end of tibia	Extends thigh
Semitendinosus	Ischium	Proximal end of tibia	Flexes lower leg
Soleus	Head of fibula	Tendon of Achilles	Extends foot
Tensor fasciae latae	Ilium	Fascia lata	Tenses fascia lata
Tenuissimus	2nd caudal vertebrae	Patella	Abducts thigh
Tibialis anterior	Tibia and head of fibula	1st metatarsal	Flexes foot
Vastus lateralis	Femur	Patella	Extends lower leg
Vastus medialis	Femur	Patella	Extends lower leg

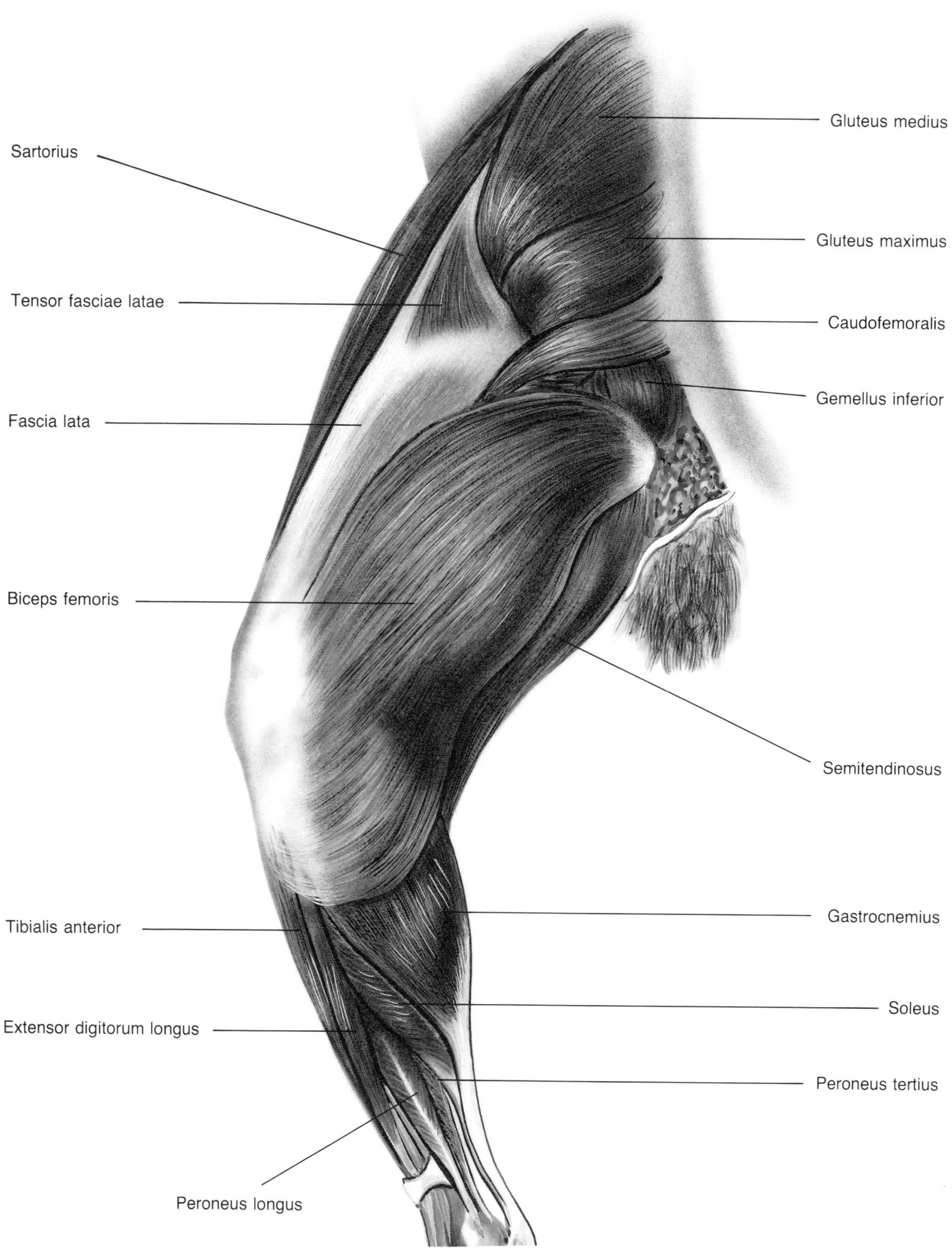

Figure 25. Left hind limb muscles, superficial (lateral view).

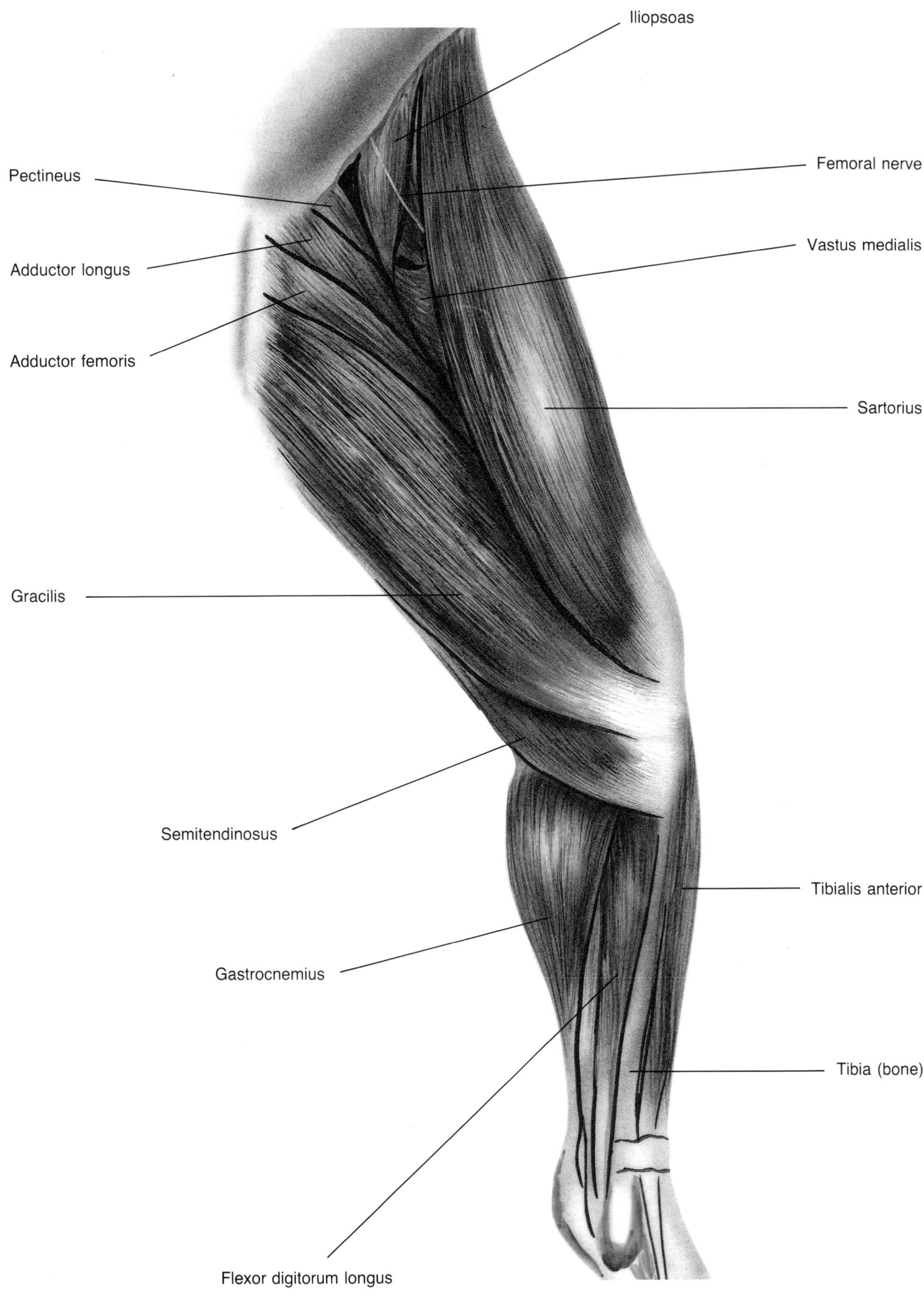

Figure 26. Left hind limb muscles, superficial (medial view).

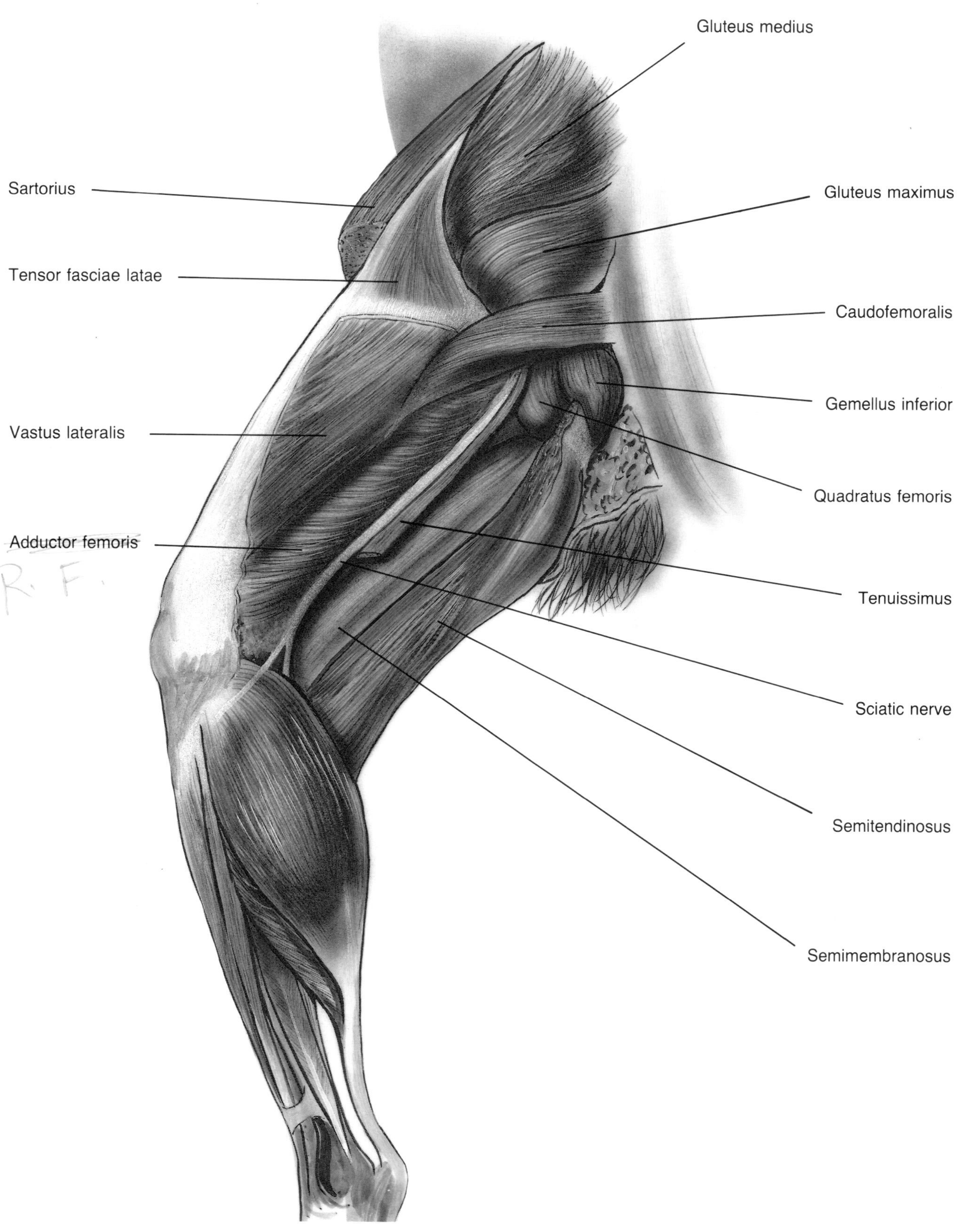

Figure 27. Left hind limb muscles, deep (lateral view).

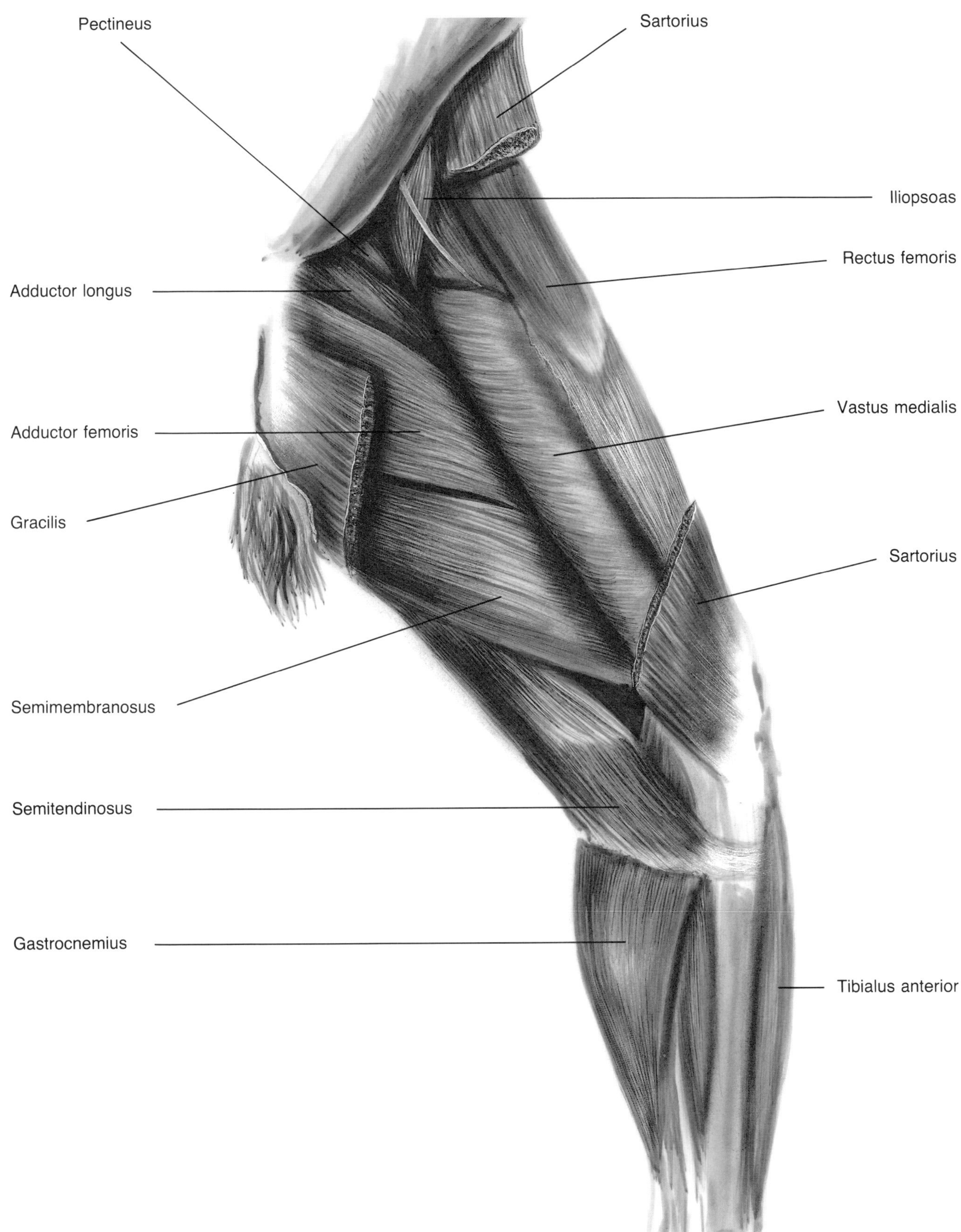

Figure 28. Left hind limb muscles, deep (medial view).

Table 5
Shoulder and Trunk Muscles

Muscle	Origin	Insertion	Action
Acromiotrapezius	Dorsal midline	Spine of scapula	Pulls scapula toward spine
External oblique abdominis	Ribs and lumbodorsal fascia	Ventral midline	Compresses abdomen
Infraspinatus	Infraspinous fossa of scapula	Greater tuberosity of humerus	Rotates humerus
Internal oblique abdominis	Lumbodorsal fascia, ilium	Midventral line	Constricts abdomen
Latissimus dorsi	4th thoracic to 6th lumbar vertebra	Humerus	Pulls arm caudo-dorsally
Levator scapulae ventralis	Atlas and occipital	Metacromion and infraspinatus fossa of scapula	Pulls scapula anteriorly
Spinodeltoid	Spine of scapula	Humerus	Flexes humerus
Spinotrapezius	Thoracic vertebrae	Spine of scapula	Pulls scapula dorsally and posteriorly
Supraspinatus	Supraspinous fossa of scapula	Greater tuberosity of humerus	Extends humerus
Transverse abdominis	Ribs, lumbar vertebrae, and ilium	Ventral midline	Constricts abdomen

Shoulder and trunk muscles *(Table 5 and Figure 29).* The abdominal muscles (except for the rectus abdominis) are shown here. The direction of the muscle fibers is the most reliable method of identification.

Questions and Activities

1. Use rubber bands and wood to make a model of a joint.
2. Describe a simple muscular action (e.g. lifting a forelimb) of a cat in terms of the major muscles used.
3. Choose a muscle and describe the impairment a cat might suffer if that muscle were paralyzed. What other muscles might compensate for the impairment?

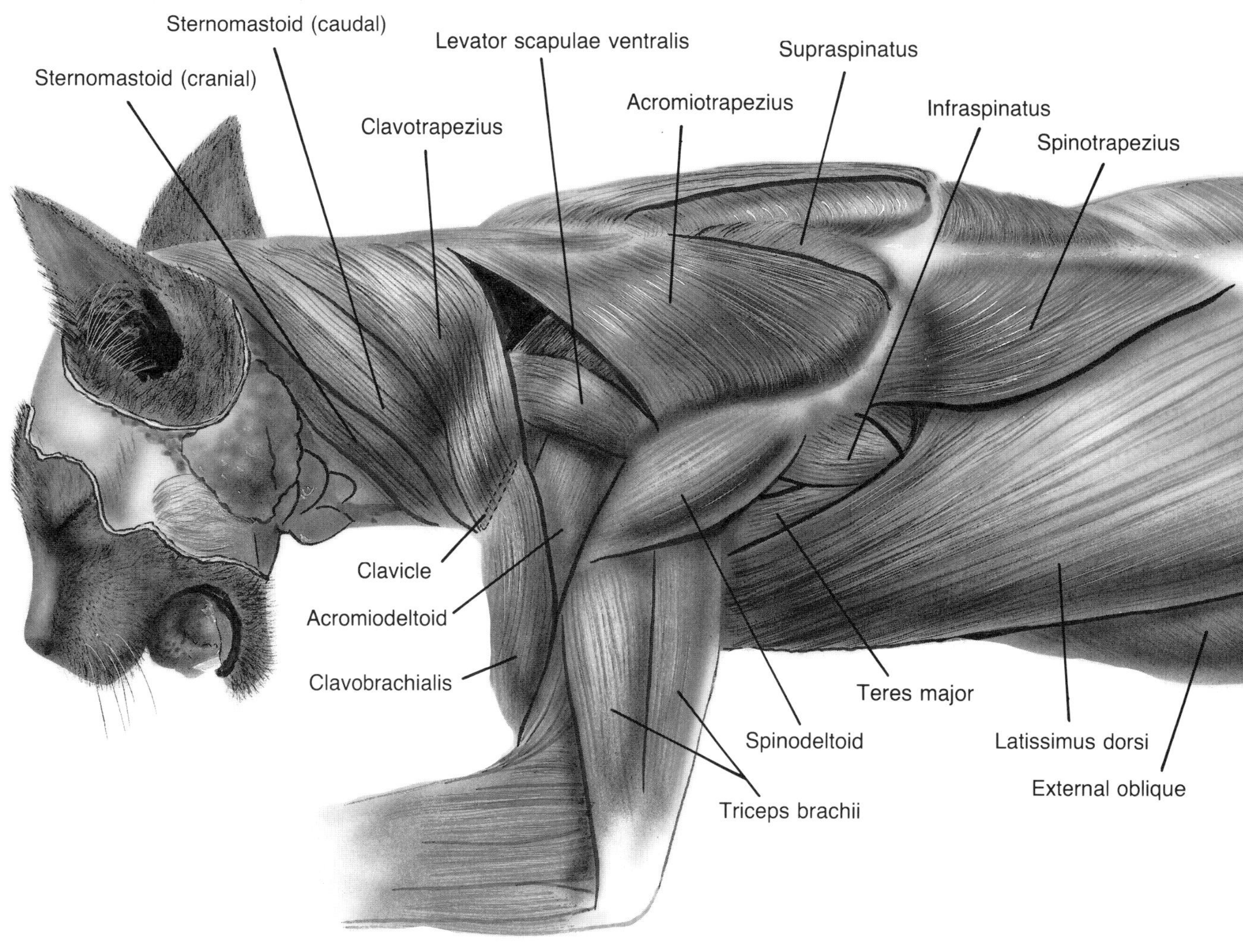

Figure 29. Shoulder and trunk muscles, superficial (lateral view).

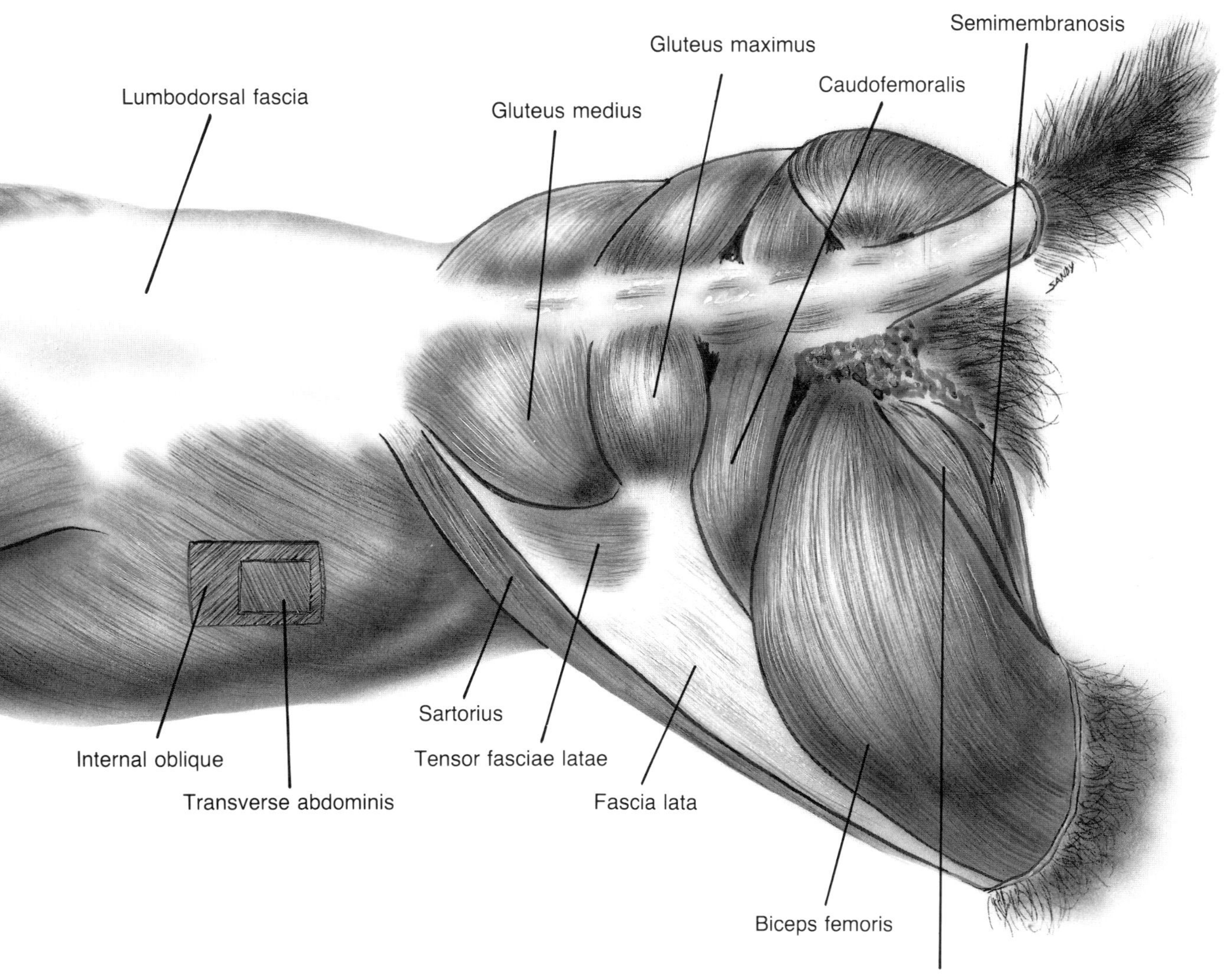

Lumbodorsal fascia
Gluteus medius
Gluteus maximus
Caudofemoralis
Semimembranosis
Internal oblique
Transverse abdominis
Sartorius
Tensor fasciae latae
Fascia lata
Biceps femoris
Semitendinosis

NOTES

Unit 4
THE DIGESTIVE SYSTEM

The mammalian body has two major cavities, the abdominal and thoracic, which contain all the internal organs exclusive of the brain. These two cavities are divided by a sheet of muscle, the diaphragm. Mammals are the only vertebrates with a muscular division between thorax and abdomen.

To examine the viscera, we must first open the two body cavities. Begin by opening the abdominal cavity. To do this, locate the posterior end of the sternum. Use forceps to lift the abdominal muscle just posterior to the sternum and cut through the muscle and into the abdominal cavity. Insert the blunt tip of the scissors into the cut and make an incision to one side of the midventral line to the posterior end of the abdominal cavity. (See Figure 30.) Beginning at the anterior end of the incision, make lateral cuts following the posterior margin of the ribs. Also make lateral cuts through the body wall at the posterior end of the abdominal cavity. (Male cats have cords running from the posterior abdomen along the surface of the thigh muscles to the scrotum. Leave these undamaged.) You can now fold back the lateral body wall and either pin it out of the way or trim it off.

If you find a dark brown substance coating the abdominal viscera, blood has leaked into the abdomen and must be flushed out. Rinse the abdomen carefully in water for several minutes until the viscera are clean.

Open the thoracic cavity as follows: Use heavy scissors or bone-cutting forceps to cut through the ribs about one cm to either side of the sternum. When all the ribs are cut, lift the sternum, and hold it in place while cutting the underlying membranes until the sternum comes free. Either cut the ribs along the sides of the chest or break them to expose the thoracic viscera.

We are now ready to begin examination of the digestive system. Use bone-cutting forceps to cut the bones at the corners of the jaw until a clear view can be had of the interior of the mouth (Fig. 31).

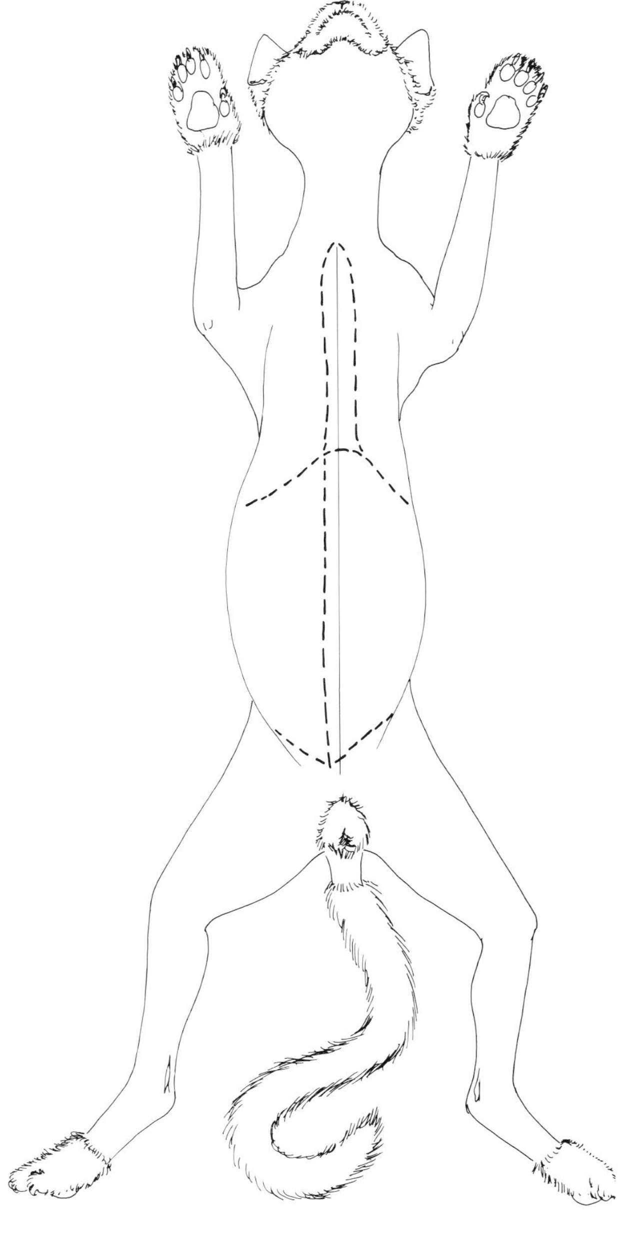

Figure 30. Diagram for opening the body cavity.

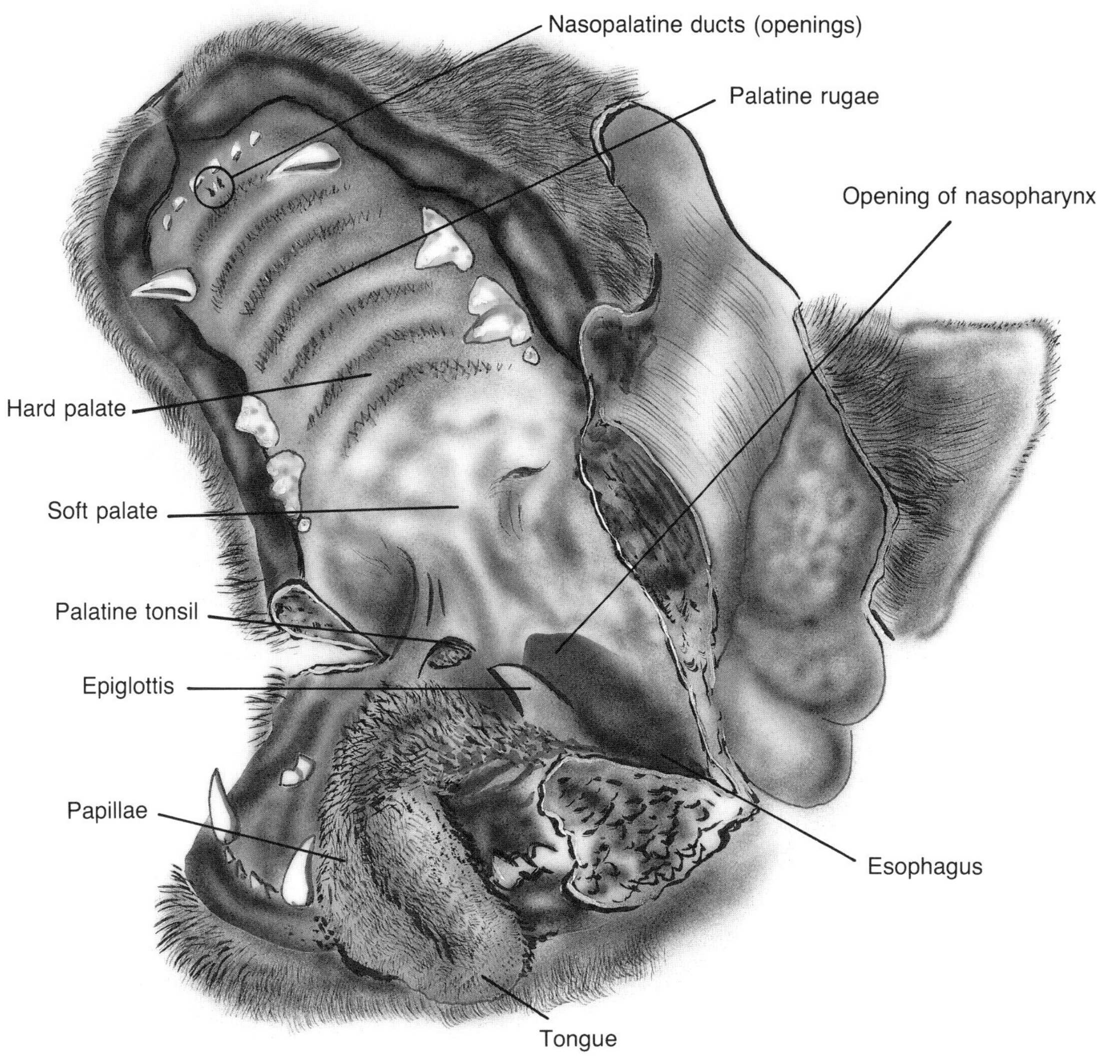

Figure 31. Dissection of the oral cavity.

Digestion begins here with the cutting of food into chunks that can be swallowed. The three major salivary glands, the *sublingual, submaxillary,* and *parotid* were exposed when dissecting the neck muscles. Review these at this time. There are two other salivary glands, the *molar* and *infraorbital.* As mentioned earlier, these are small and difficult to find. The molar gland is located in the corner of the mouth just under the skin and is somewhat diffuse. Do not look for the infraorbital gland as it is located in the orbit and you would have to remove an eye. All the salivary glands empty by ducts into the mouth, where their secretions (saliva) lubricate the food. Saliva also contains enzymes that begin chemical digestion of the food. The rough *papillae* (of which there are several types) of the tongue rasp meat off bones. The ridges help grip the food. Just posterior to the incisors are the openings of the two *nasopalatine ducts.* These communicate with the nasal cavity through the incisive foramens (see Fig. 4). The palate or roof of the mouth is divided into two parts, the *hard palate,* which is underlain by bone (compare Figs. 31 and 4), and the *soft palate* at the back of the mouth, which is not underlain by bone. Near the corners of the mouth are the *palatine tonsils,* parts of the lymphatic system. As food is swallowed, the *epiglottis* is depressed, closing the *glottis,* the entrance into the larynx. This prevents food from entering the respiratory tract.

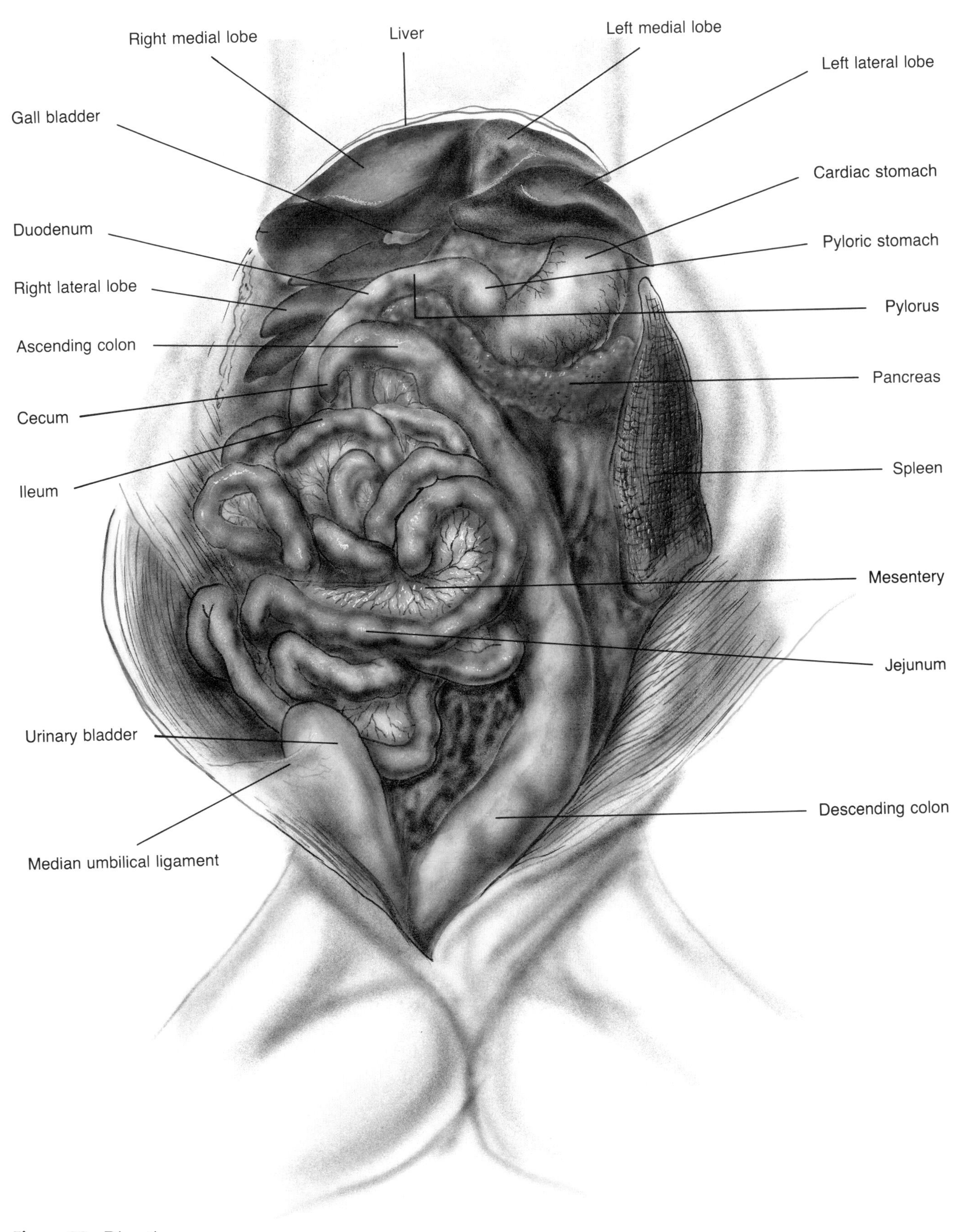

Figure 32. Digestive organs.

The anterior opening of the *esophagus* can be seen, but most of this tube is covered and will be visible only later in your dissection. Now turn your attention to the abdominal digestive organs. Covering these organs is a membrane, the *greater omentum*. Carefully remove the omentum and locate the organs and features shown in Figure 32. The urinary bladder and spleen are not digestive organs. The *spleen* is an important part of the lymphatic system and the urinary bladder is part of the urogenital system. Perhaps the most obvious feature of the digestive system is its marked asymmetry. This results from the fact that the digestive system is coiled to fit into a body that is much shorter than the digestive system. An herbivore has a proportionately longer digestive system than does a carnivore.

The *stomach* is a curved, muscular sac. The anterior, concave surface is called the *lesser curvature* while the posterior, convex surface is called the *greater curvature*. The esophagus connects the mouth and stomach. Food is moved down the esophagus by waves of muscular contraction referred to as peristalsis. The opening from the esophagus into the stomach is closed by the *cardiac sphincter*. The connection of the esophagus to the stomach is not visible at this time. (This connection can be established after dissecting the respiratory system.)

The larger, anterior portion of the stomach to which the esophagus attaches is the *cardiac* region. The small intestine attaches to the *pyloric* end of the stomach and the *pyloric sphincter* controls the release of food from the stomach into the small intestine. Gastric glands in the stomach wall secrete a highly acidic digestive fluid that contains pepsin, a protein-digesting enzyme. Muscular contractions of the stomach wall mix the food and gastric juices and by the time the food is released from the stomach, it has become a semisolid material called chyme.

The *liver* and *pancreas* are two large glands associated with the digestive system. Bile, a digestive juice secreted by the liver, is stored in the *gall bladder*. The *cystic duct,* which drains bile from the gall bladder, is joined by the *hepatic duct* from the liver to form the *common bile duct,* which enters the small intestine. Trace the pathways of these ducts, removing membranes and portions of the liver lobes as necessary.

Bile contains enzymes that aid digestion of starches, fats, and proteins. Bile, which is alkaline, also helps neutralize the acidic chyme as it enters the small intestine. The liver has many functions in addition to the production of bile. In fact, it is the site for so many biochemical reactions that it is sometimes referred to as the body's "chemical factory." The liver is incompletely divided into right and left lobes; each lobe is further divided into medial and lateral lobes. The *right lateral lobe* is again subdivided into cranial and caudal portions. There is also a small division of the liver, the *caudate lobe,* which is not shown in Figure 32. To find this lobe, lift the left side of the liver and look dorsally.

The pancreas is a two-lobed gland. One lobe, the *gastrosplenic,* lies near the greater curvature of the stomach. The second lobe, the *duodenal,* lies along the upper section of the duodenum. The pancreas secretes juices containing enzymes that aid in the digestion of starches, fats, and proteins. The pancreas is also an endocrine gland, producing insulin, which is involved in the regulation of blood glucose levels.

The first part of the small intestine is the *duodenum,* which curves to the right and posteriorly. The duodenum then curves to the left and turns posteriorly again. This posterior turn marks the beginning of the *jejunum,* the second part of the small intestine. The *ileum* is the third and final part of the small intestine. There is no clear demarcation between the jejunum and the ileum. The jejunum is proximal to the duodenum and the ileum is proximal to the large intestine. There is a small blind sac of the large intestine, the *cecum,* which extends past the junction of the ileum and large intestine. This is a homologue of the human appendix. The small intestine is the chief organ of chemical digestion and the chief organ of nutrient absorption. Water and some vitamins are absorbed in the large intestine.

Like the small intestine, the large intestine has three indistinct sections. These are the *ascending colon,* to which the ileum attaches, the *transverse colon,* and the *descending colon.* The ascending colon runs anteriorly along the right. The short transverse colon passes to the left, and the descending colon runs posteriorly to the *rectum,* the terminal portion of the digestive tract. In Figure 32 the large intestine has been pulled to the left, changing these relationships. Also, the intestines tend to move about in the living animal.

Questions and Activities

1. You have now observed at least one other characteristic unique to mammals. What is it?
2. Cut open and wash out a short section of the small intestine and observe the villi under a stereomicroscope.
3. Cut open the stomach and describe its internal appearance.

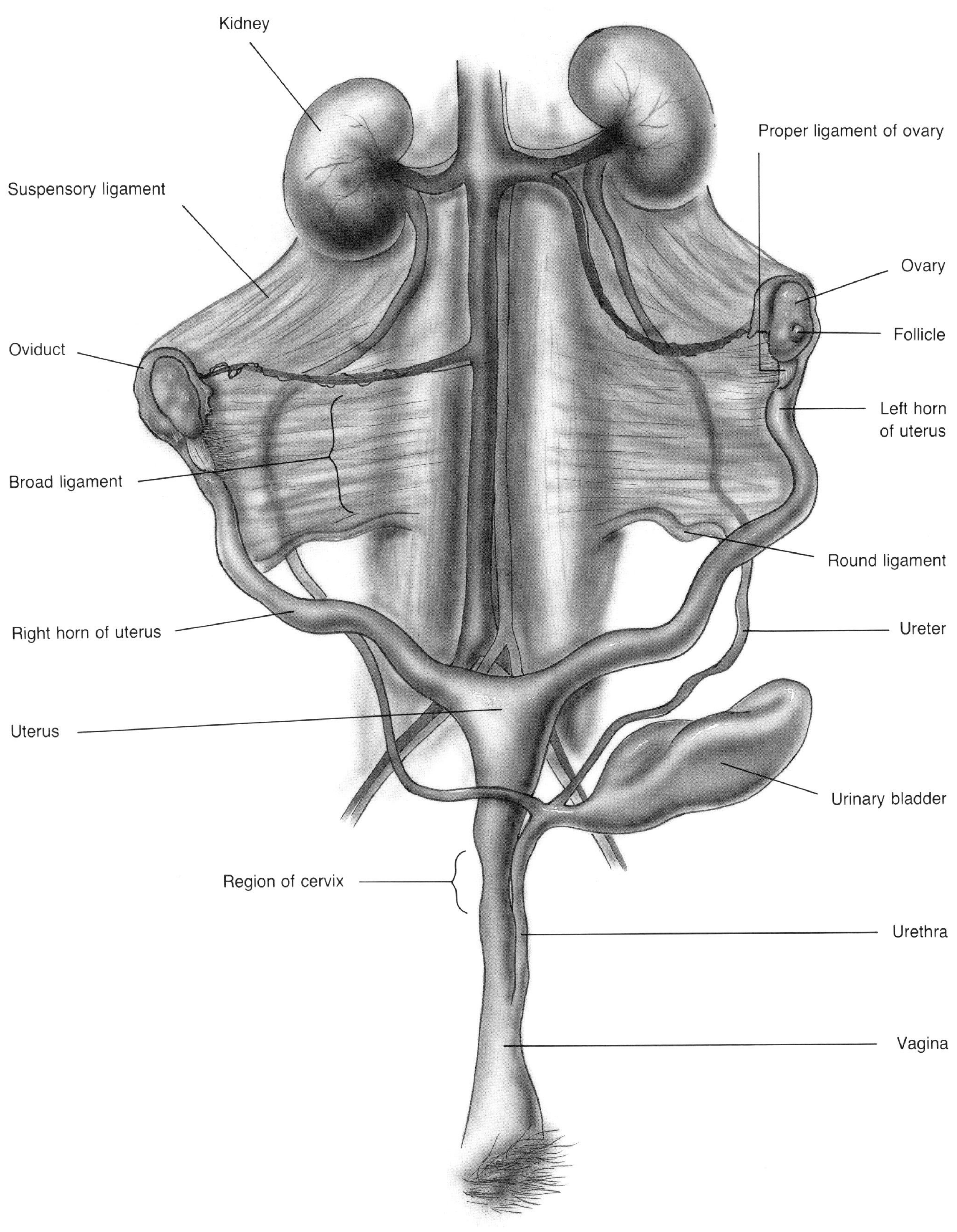

Figure 33. Female urogenital system.

Unit 5
THE UROGENITAL SYSTEM

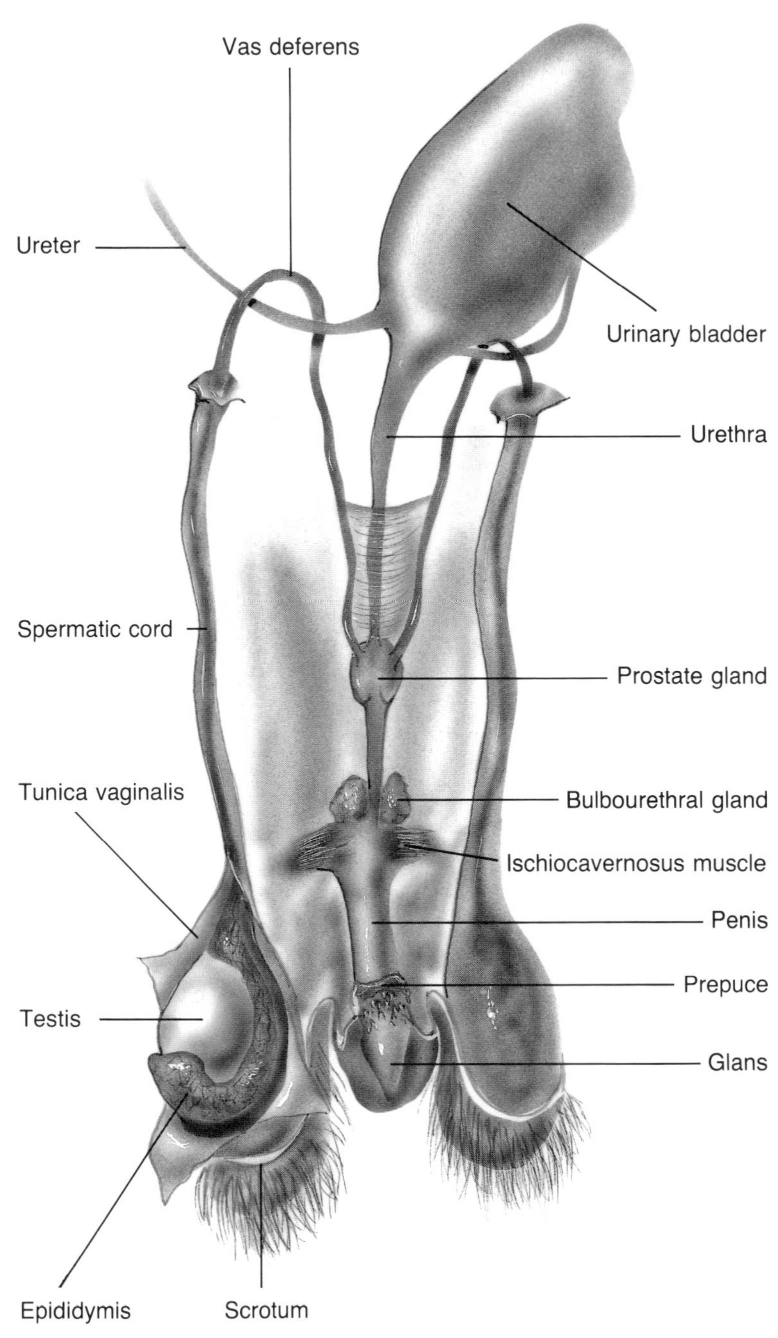

Figure 34. Male urogenital system.

Push the intestines to one side until the *kidneys* are located (Fig. 33). The kidneys are dorsal to the *peritoneum,* the membrane that lines the inner body wall and coats the viscera. The kidneys are usually surrounded by a cushioning layer of fat. Both fat and peritoneum must be removed to obtain a clear view of the kidneys. The medial surface of each kidney has an indentation called the *hilus,* to which are attached the renal blood vessels and the *ureter.* Follow the ureters to their attachment to the *urinary bladder.* To locate the attachment site, use bone-cutting forceps to cut through and remove the pubic bones. Once the ureters' attachments are located, remove more tissue to locate the *urethra.*

Blood brought to the kidneys is filtered to remove waste products, especially urea. Also removed from the blood are substances (such as salt) that have been ingested in excessive amounts. Thus, the kidneys help regulate the osmotic pressure of the blood. Urine excreted by the kidneys goes through the ureters to the urinary bladder for temporary storage. Contractions of the bladder wall force urine out through the urethra. These parts, kidneys to urethra, comprise the urinary system, and they are the same for male and female.

The genital system (which is closely associated with and attached to the urinary system) is, of course, different for the two sexes. If you have a male cat, refer to Figure 34 in dissecting the genital system; otherwise, continue to refer to Figure 33. Examine a dissection of both male and female genital systems.

To dissect the genital system, remove

the connective tissue around the urethra and genital parts. There are blood vessels located dorsal and lateral to the rectum (these are shown for reference in Figure 33), and you should be careful not to damage them.

Female Genital System

The *ovaries* are located just posterior to each kidney. The *suspensory, broad,* and *round ligaments* hold the ovary and uterine horns in place. The ovary is held to the *uterine horn* by the *proper ligament of the ovary*. A mature *follicle* is shown in the drawing of the left ovary. Follicles can also be seen by cutting open the ovary. Each follicle contains a developing ovum. The ovum is eventually expelled from the follicle and passes into the *oviduct*. Notice that there is no direct connection between the ovary and the oviduct; rather, the oviduct has an opening that is pressed close against the ovary. If your specimen is pregnant, the uterine horns will be enlarged (perhaps greatly) over the condition shown in Figure 33. In this case, there will be enlargements along the uterine horns and each enlargement marks the position of a fetus. Cut open the uterus to examine the fetuses. One female cat can produce three to four litters of up to six kittens per year.

Male Genital System

Remove tissue ventral to the urethra to find the junction of the urethra and the *vas deferens*. At this junction should be a swelling, the *prostate gland*. The vas deferens carries sperm from the *testes*. The prostate gland adds fluid secretions to the sperm, producing the semen. During copulation, muscular contractions of the prostate gland force semen down the urethra through the *penis*. Along the way, more fluid secretions are added by the *bulbourethral glands*. The bulbourethral and prostate glands may be small and difficult to locate in young, sexually immature specimens. Near the bulbourethral glands are the *ischiocavernosus muscles,* which help restrict blood flow from the penis. When these muscles contract, sinuses within the penis fill with blood, erecting the penis in preparation for copulation. Slit the *prepuce* (foreskin) to locate the *glans penis*. The glans penis contains nerve endings that are stimulated during copulation, leading to ejaculation of the semen. Find the *urogenital opening* at the tip of the penis.

Trace the vas deferens from the prostate gland to the abdominal wall. The vas deferens passes through the wall by way of the *inguinal canal*. In the inguinal canal, the vas deferens is covered by a sheath and is joined by spermatic blood vessels to form the *spermatic cord*. Trace this cord into the scrotum. There the outer sheath of the spermatic cord, the *tunica vaginalis,* encloses the testis. Carefully slit the tunica to reveal the testis and *epididymis*.

Questions and Activities

1. Diagram and name the major types of mammalian uteri.
2. The male and female genital systems originate from the same embryonic tissues. List the parts of the male genital system and give the female homologues.

Unit 6
THE CIRCULATORY SYSTEM

Anterior Blood Vessels

The *heart* occupies the center of the thoracic cavity and is covered by a tough membrane, the *pericardium*. In the pericardium is the *pericardial vein*. Remove the pericardium and locate the *anterior vena cava,* the large vein attached to the anterior side of the heart. Follow this vein and its branches, locating the veins shown in Figure 35.

Take careful note of the areas drained by each vein. The *internal jugular veins* drain the brain and spinal cord.

The anterior arteries are generally parallel and dorsal to the veins. Once the veins have been clearly observed, remove them, beginning at the heart. (As you do so, try not to damage the nearby nerves, especially the phrenic. See Figure 45.) Then locate the arteries shown in Figure 36. Note which areas are supplied with blood by which arteries. The *internal carotid arteries* supply the brain. The *linguals,* as their name implies, go to the tongue and also to the pharyngeal muscles.

Heart

Lift the heart and cut the vessels and connective tissue attaching the heart dorsally. (Again, do as little damage to nerves as possible.) Identify the parts shown in Figures 37 and 38. The right ventricle of the heart pumps blood to the pulmonary (respiratory) circulation and the left ventricle to the general or systemic circulation. The *arterial ligament* is the remnant of the *ductus arteriosus* which, during fetal life, shunts blood from pulmonary to systemic circulation. The *auricles* are sac-like flaps of the *atria*. The atria receive blood returning to the heart. The *ventricles* are the main pumping chambers. The dorsal view shows the *posterior vena cava,* which returns blood from regions posterior to the heart. Also visible are the *pulmonary veins and arteries.* Trace the connections of these to the lungs.

Open the right side of the heart by making the cuts shown by the dashed lines on Figure 39. Locate the structures shown on Figure 40. The *papillary muscles* are attached to the flaps of the *tricuspid valve* by the *chordae tendinae* and prevent the valve from being pushed into the atrium when the ventricle contracts. Blood from the coronary veins returns to the right atrium through the opening of the *coronary sinus.* Open the left side of the heart as shown by the dashed lines and arrows of Figure 41. Locate the structures shown in Figure 42.

Abdominal Blood Vessels

Make an incision near the base of each thigh just lateral to the rectum. Remove the rectum intact and push the digestive tract to one side, cutting membranes as necessary to achieve a good view of the dorsal wall of the abdomen. Remove fat and tissue to locate the abdominal vessels shown in Figure 43. The *adrenal gland* shown in this figure is an important endocrine gland.

Notice that the *aorta,* which at first is dorsal to the posterior vena cava, comes to lie ventral to it.

Usually a capillary bed gives rise to a vein that joins with other veins, eventually to terminate at the heart; however, a capillary bed may also give rise to a vein that terminates in another capillary bed. These are called portal systems, and the largest is the *hepatic portal system,* which carries blood from capillary beds in the intestines to capillary beds in the liver. If yours is a triple-injected cat, the hepatic portal system has been injected with yellow latex. Figure 44a was made by spreading out the small intestines and removing large portions of the pancreas. The intestines and blood vessels are held in place by the *mesenteries.* Also held in the mesenteries are lymph nodes, which may have to be removed before a clear view of the blood vessels can be obtained. Figure 44b was made by rotating the small

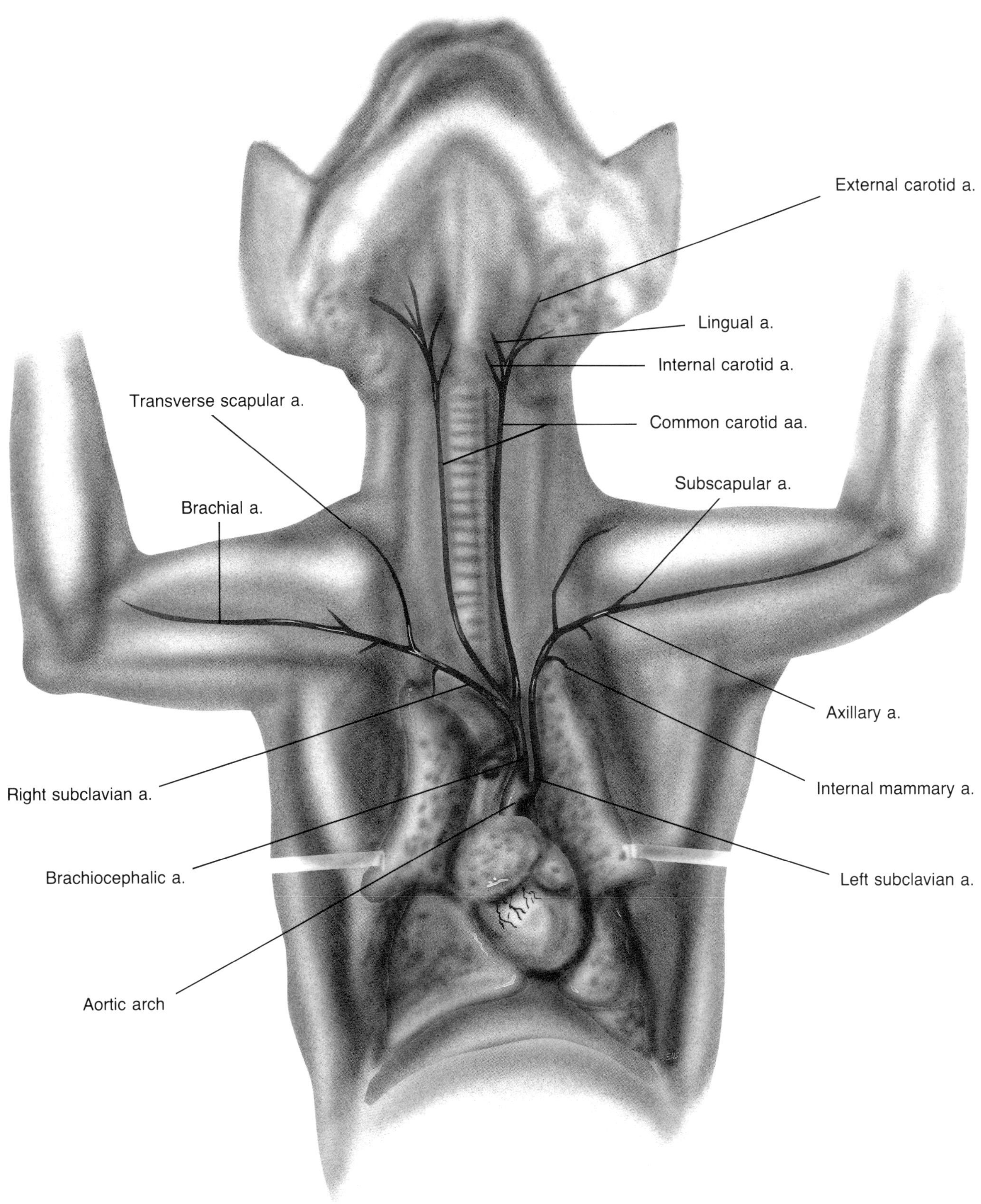

Figure 35. Veins anterior to the heart.

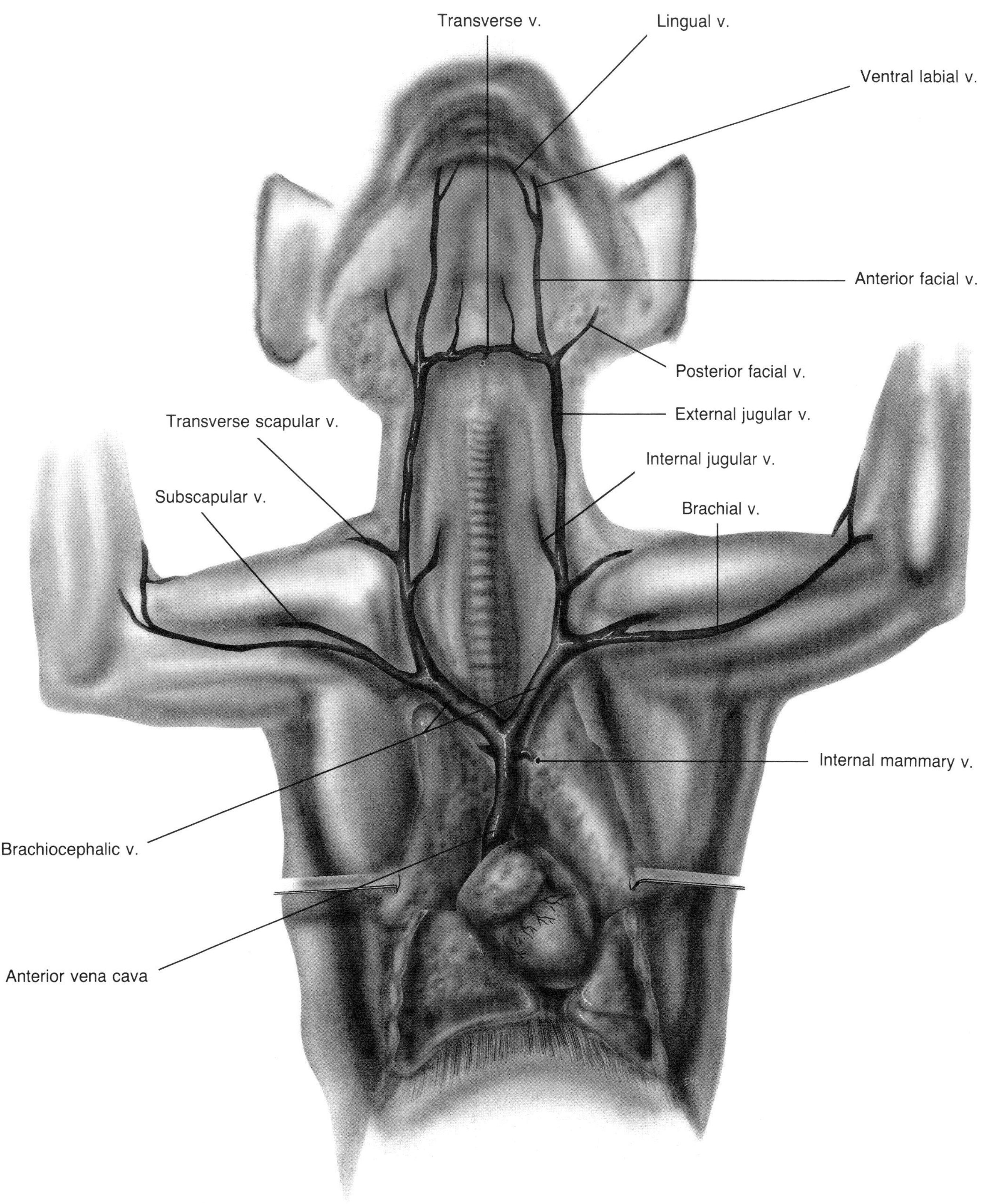

Figure 36. Arteries anterior to the heart.

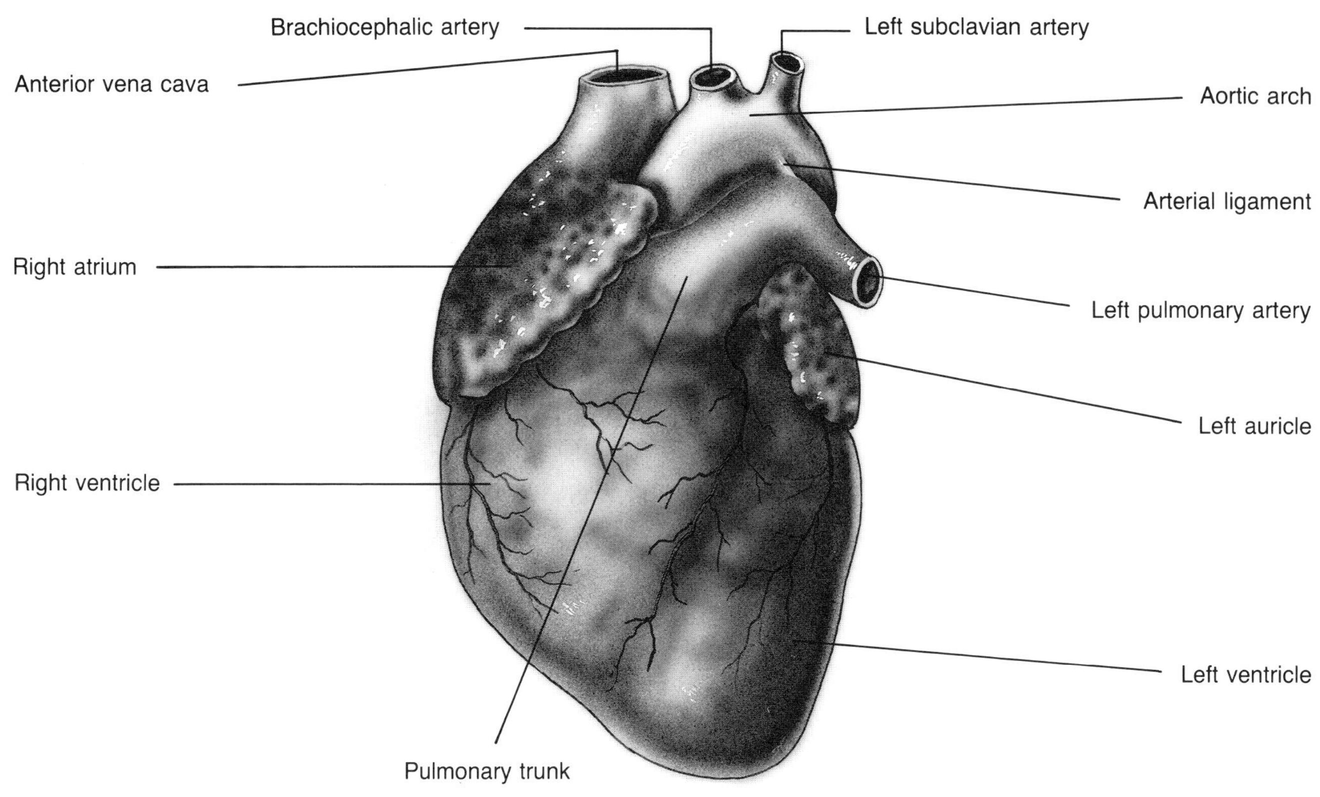

Figure 37. Heart (ventral view).

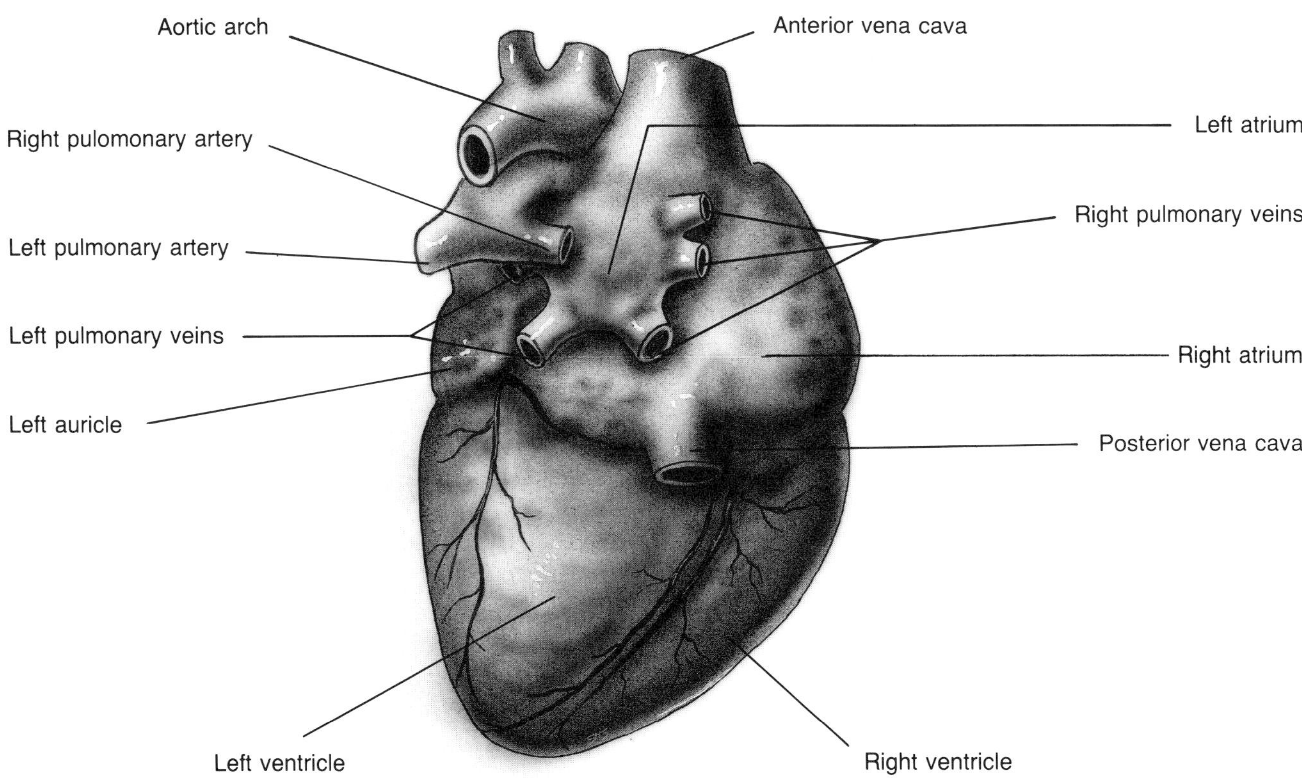

Figure 38. Heart (dorsal view).

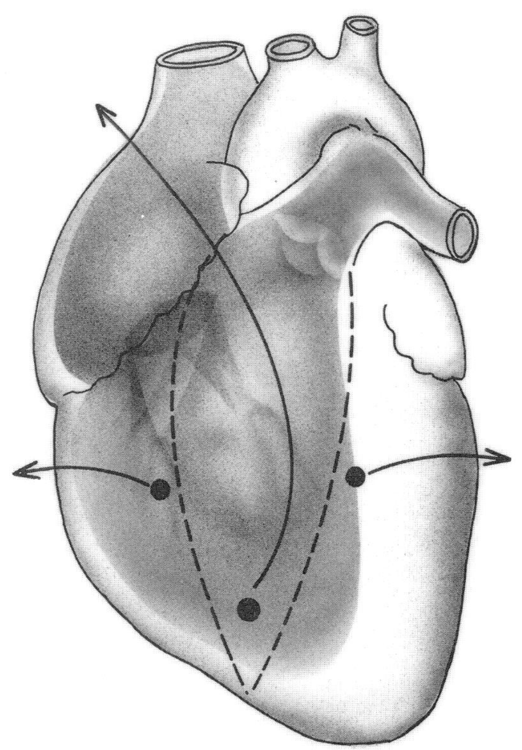

Figure 39. Dissection diagram of right side of heart.

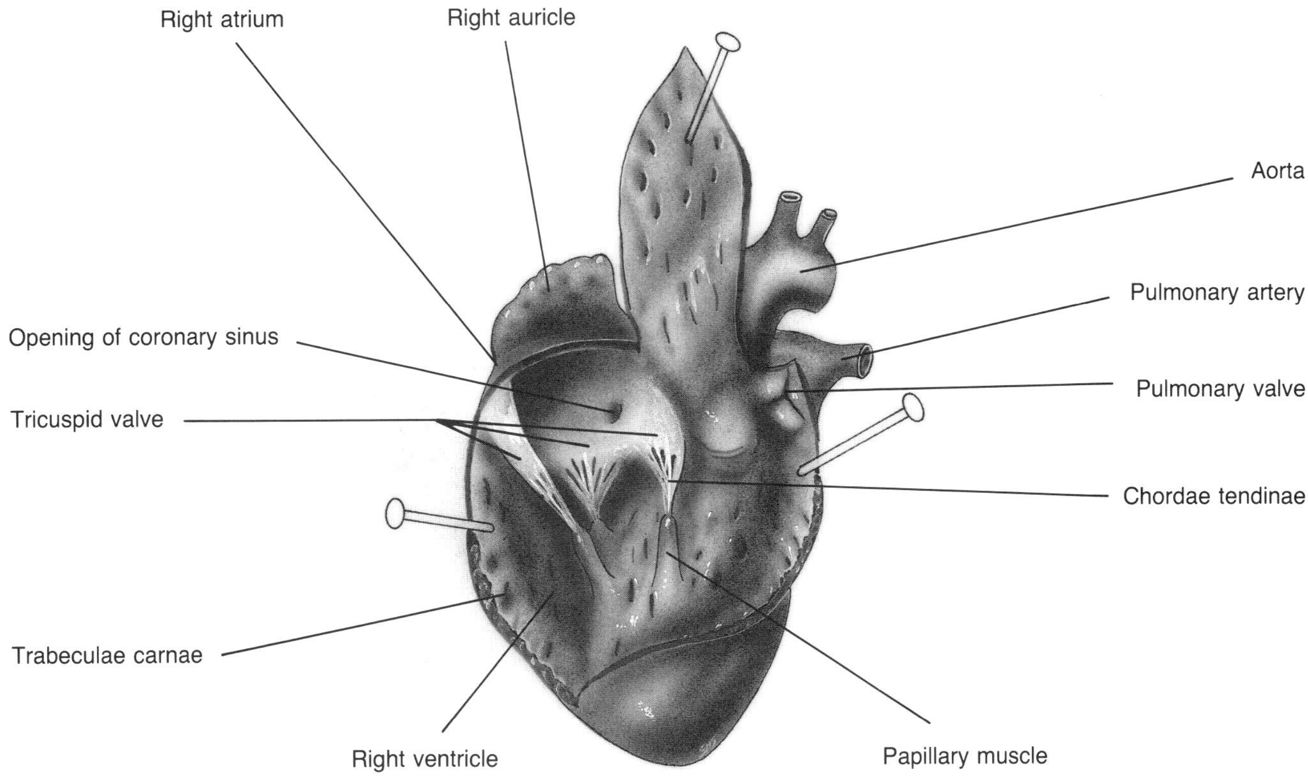

Figure 40. Right ventricle and atrium of the heart.

intestines clockwise. Parts of the liver have been scraped away to show the branching of the *portal vein* as it enters the liver. Slice through the liver to find strands of red, blue, and yellow latex. The blue indicates veins that drain blood to the posterior vena cava.

Notice that many branches of the portal vein are accompanied by branches of the *superior mesenteric artery*. These arteries generally have the same names as the veins they accompany.

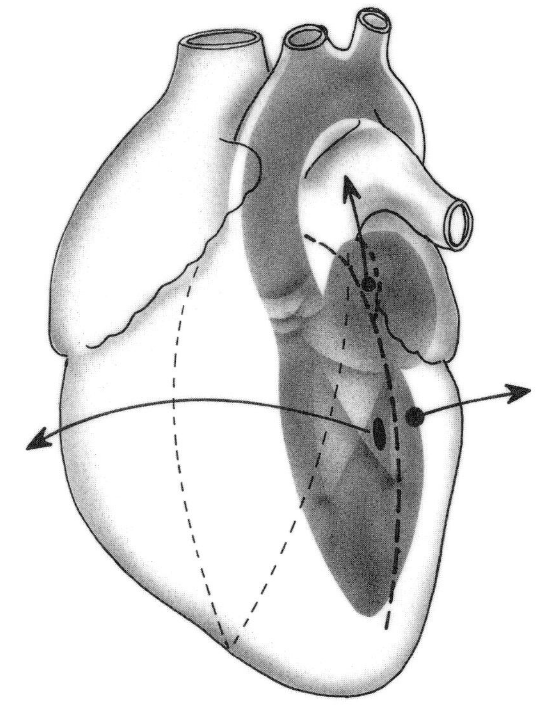

Figure 41. Dissection diagram of left side of heart.

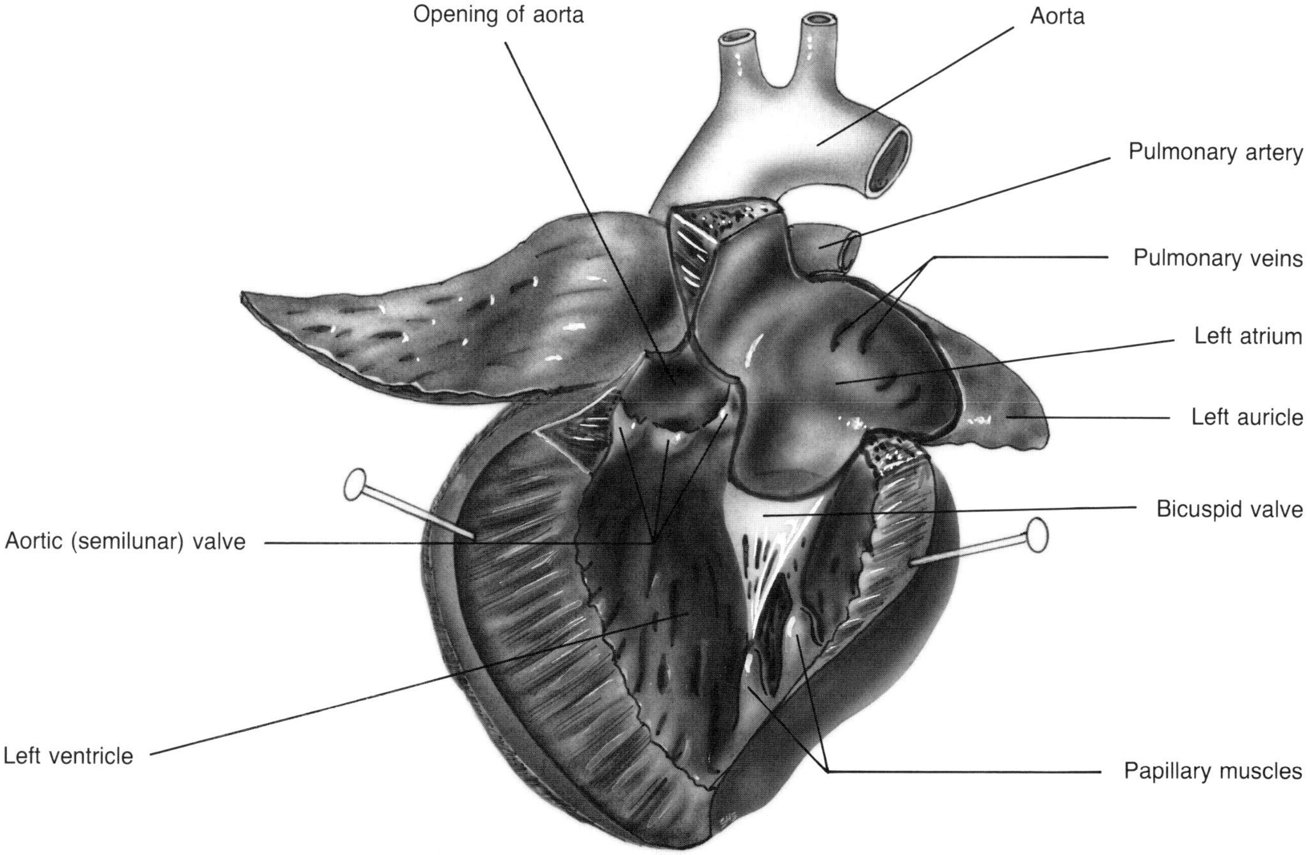

Figure 42. Left ventricle and atrium of the heart.

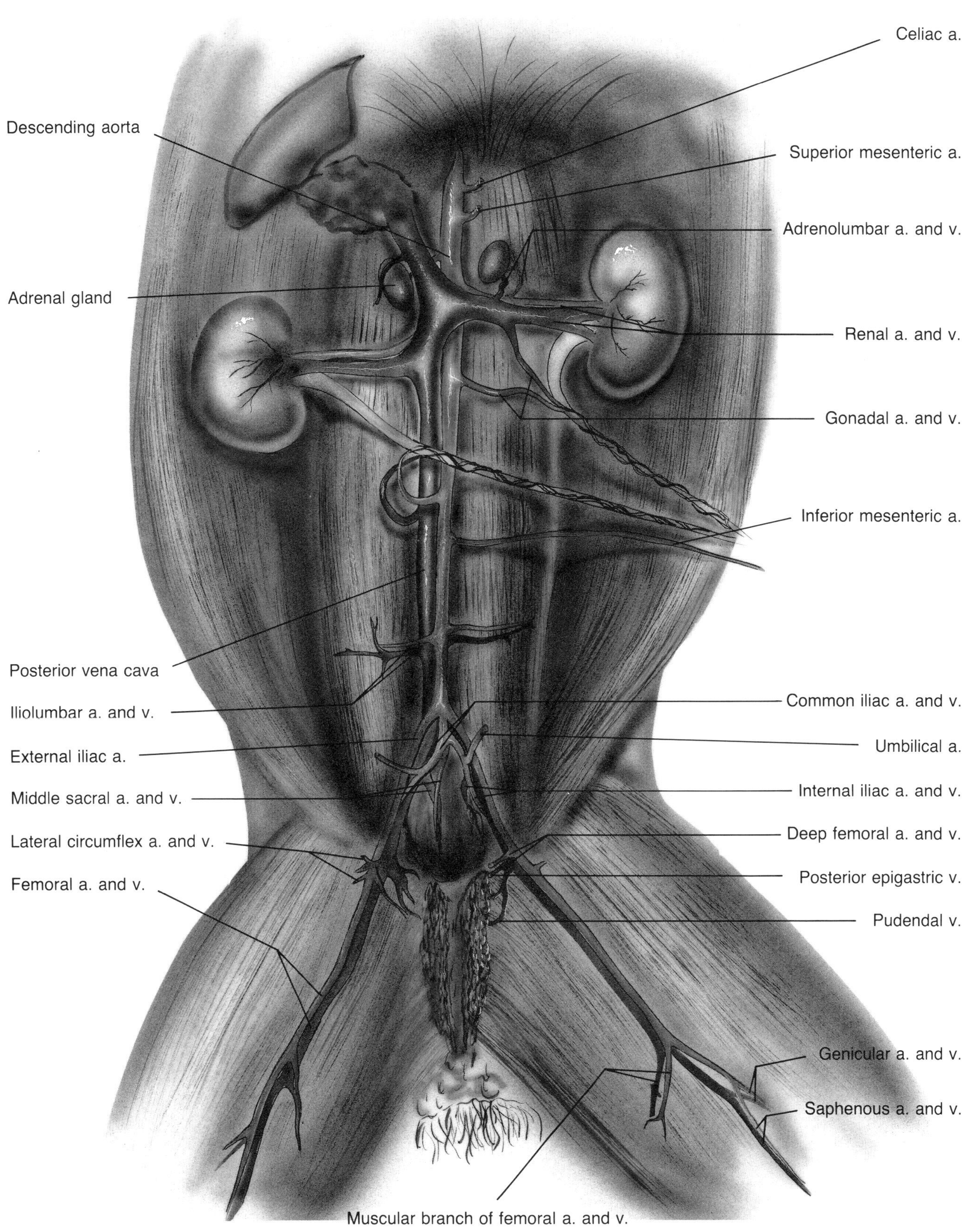

Figure 43. Veins and arteries posterior to the heart.

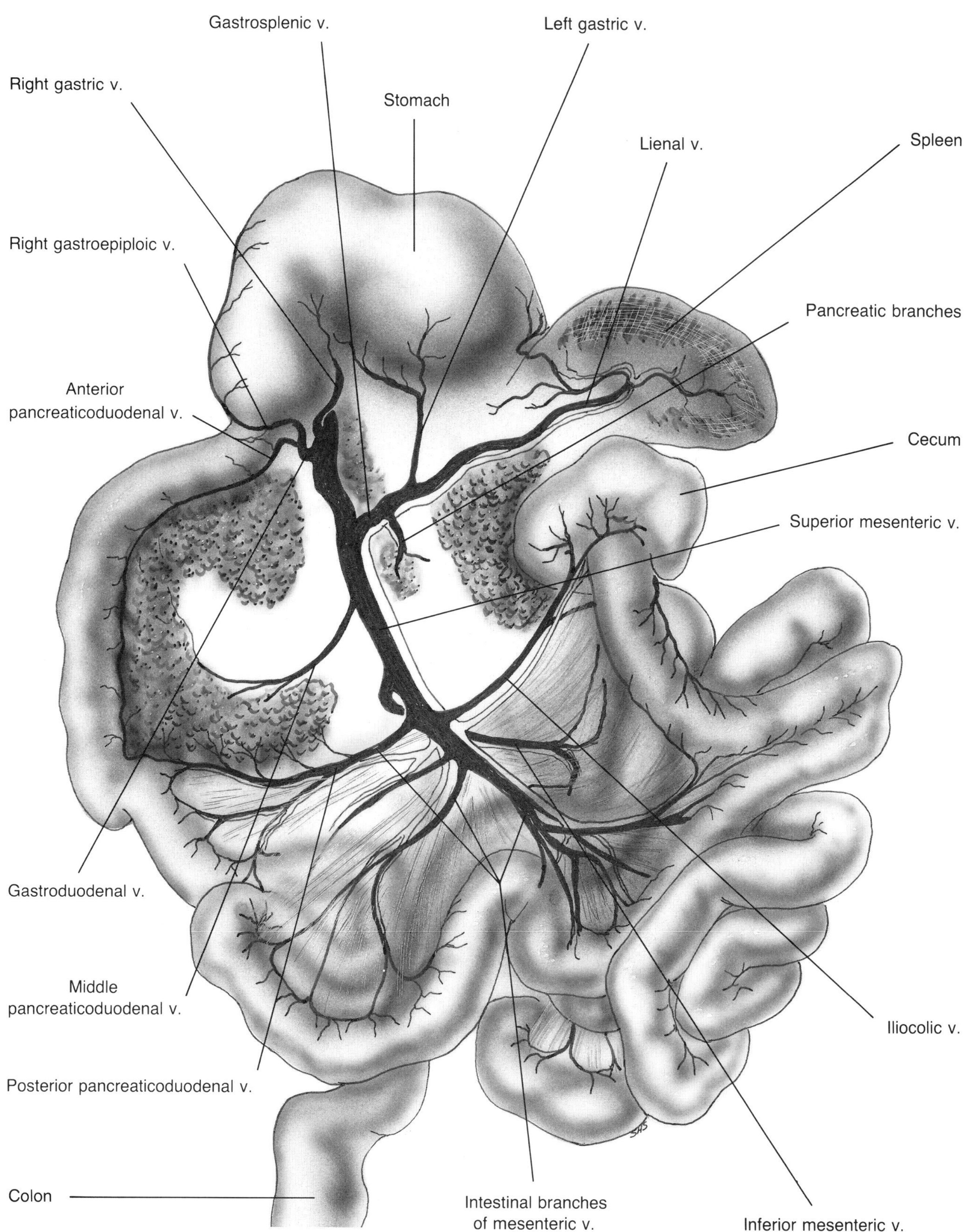

Figure 44a. Hepatic portal veins.

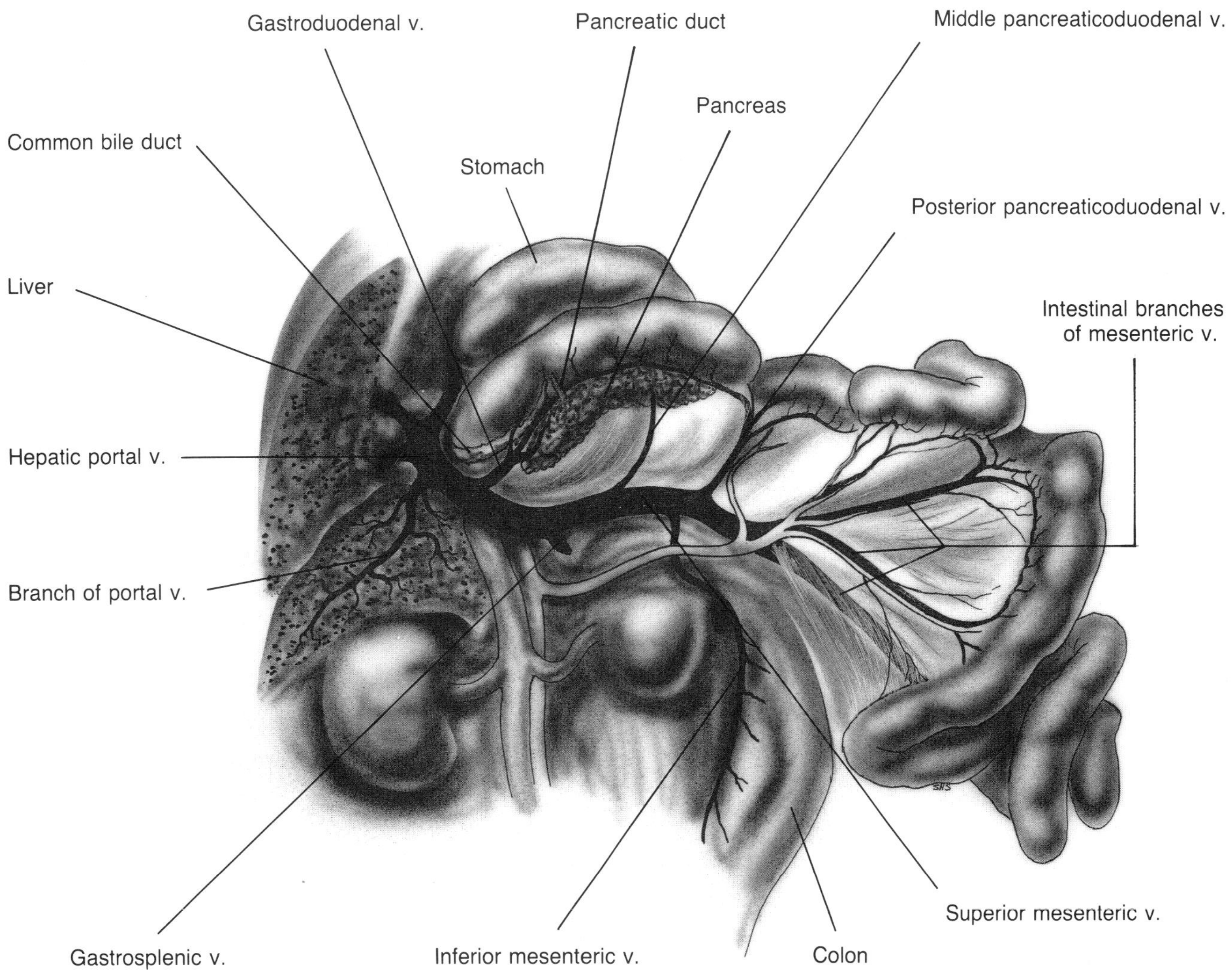

Figure 44b. Hepatic portal veins.

Questions and Activities

1. What feature(s) of the cat's circulatory system are unique to mammals?
2. Trace the path of circulation in the adult mammalian heart. How would this differ in the fetus?
3. Describe any "abnormalities" of the circulatory system found by yourself or other students in your lab section.

53

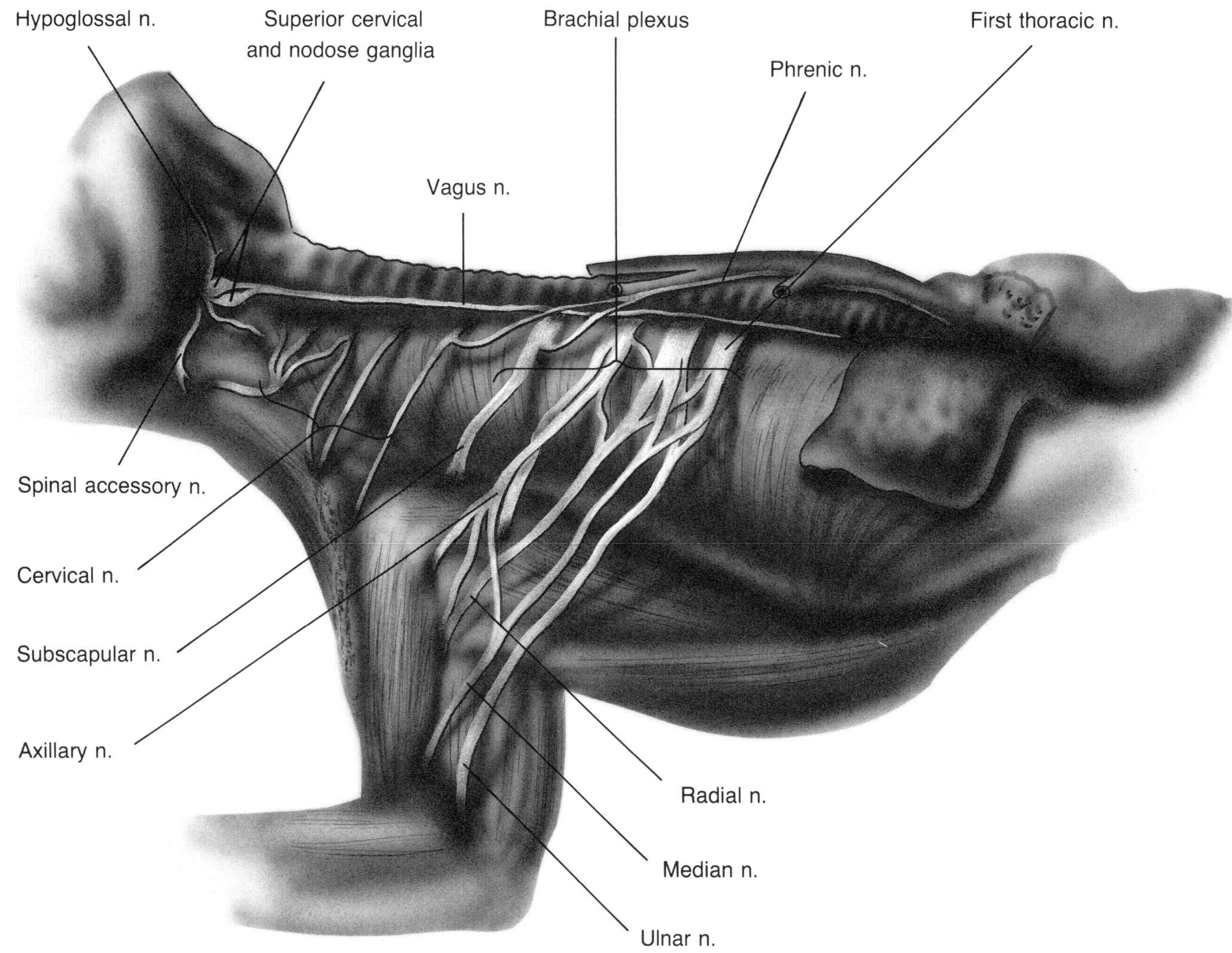

Figure 45. Nerves of the neck and brachial plexus.

Unit 7
THE NERVES OF THE
CERVICAL AND BRACHIAL REGIONS

The cat has twelve pairs of cranial nerves (nerves that originate directly from the brain) and 38 pairs of spinal nerves. Of these spinal nerves, eight pairs originate in the neck region and are referred to as cervical nerves, and thirteen pairs originate in the thorax and are called thoracic nerves. Three cranial nerves (hypoglossal, vagus, and spinal accessory) can be found in the neck. Additionally, several *cervical nerves* and the *first thoracic nerve* can be found in the neck and upper thoracic regions.

The nerves are the same color as the fat and connective tissue that surrounds them, and many of them are delicate, so the dissection must be made carefully. Use a forceps with fine points, a teasing needle or fine-pointed probe, and fine dissection scissors to expose the nerves. A network of nerves going to the arm (brachium) should be visible. This is the *brachial plexus* (Fig. 45), and it is formed by the four posterior cervical nerves and the first thoracic nerve. These nerves are large and fairly sturdy. Begin your dissection by exposing the brachial plexus and its branches. Once this is done, the other nerves can be exposed with more confidence. Follow the *vagus nerve* forward to locate the *superior cervical* and *nodose ganglia* and the nearby cranial nerves.

For information on the cranial nerves, see *Unit 9: The Central Nervous System*. See Table 6 for the spinal nerves exposed by this dissection and the major muscles or organs that they innervate.

Table 6
Spinal Nerves and Innervations

Nerve	Innervation
Suprascapular	Supraspinatus
Subscapular	Subscapularis, teres major, latissimus dorsi
Phrenic	Diaphragm
Axillary	Spinodeltoid, long head of triceps brachii
Radial	Triceps brachii and lower forelimb
Median	Lower forearm
Ulnar	Flexor capri ulnaris

Questions and Activities

1. Extend the dissection to the backbone.
2. You have previously observed the sciatic nerve in the hind limb. Trace it and its branches.
3. Trace the brachial nerves into the forearm.

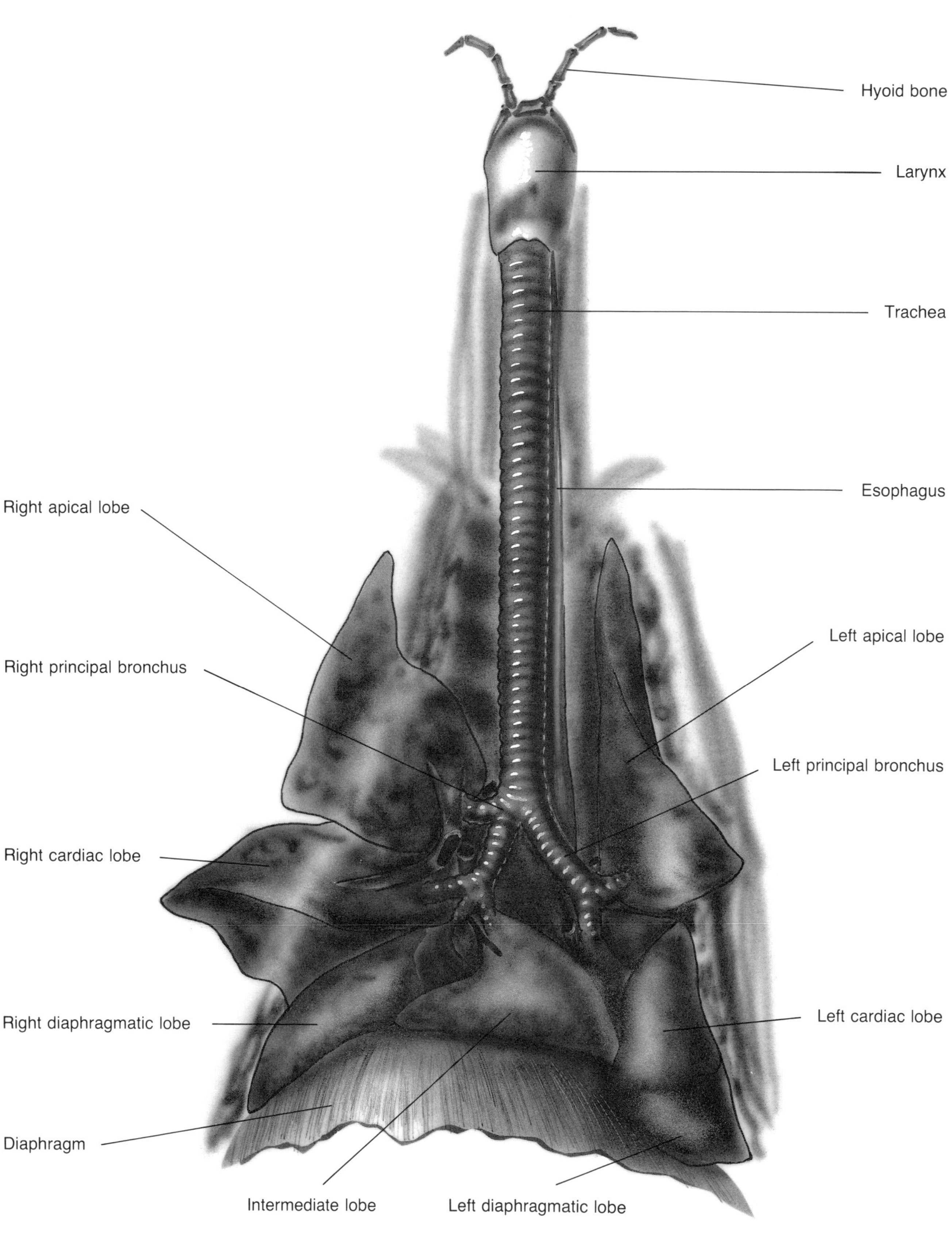

Figure 46. Respiratory system.

Unit 8
THE RESPIRATORY SYSTEM

By this time the respiratory system should be well exposed. Clear away any remaining muscle that obscures the *trachea, larynx,* and hyoid bone. Locate the structures shown in Figure 46. As the lungs inflate and deflate during breathing, they rub against the inner thoracic walls. To prevent damage from this friction, both the lungs and inner thoracic wall are lined with smooth lubricating membranes, the pleura.

Make a thin slice through the lobe of a lung and observe it under a stereomicroscope. You will observe *bronchioles* and the many blood vessels within the lung.

When the respiratory system has been thoroughly studied, expose the esophagus from the pharynx to the stomach.

Questions and Activities

1. Describe the actions that take place in breathing.
2. What would you describe as the origin(s) and insertion(s) of the diaphragm?

NOTES

Unit 9
THE CENTRAL NERVOUS SYSTEM

For this dissection, place the cat on its ventral side and use a scalpel to remove the ears and all muscle from the head down to the level of the eyes and zygomatic arch. Also remove the neck muscles to expose the cervical vertebrae and the occipital area of the skull. Try to clean the bone of all muscle. Using forceps and bone-cutting forceps, break the vertebrae and pick them away to expose the *spinal cord*. Use the bone-cutting forceps to break a hole at the back of the skull. Then carefully break away the bone from the top and sides of the cranium to expose the brain. Once the brain is well exposed, cut the spinal cord about 3 cm posterior to the brain and raise the brain from the floor of the cranium. Cut away the spinal nerves and reach under the brain with fine dissecting scissors or a needle and cut the cranial nerves. Leave as much of the cranial nerves attached to the brain as possible. Remember that there is a partial bony partition between the cerebellum and cerebrum. This may require some extra cutting before the brain will come out. Once the brain is out, examine it, locating the features shown on Figures 47 and 48. Most of the cranial nerves will probably have been lost. The olfactory nerves always break because they pass through the cribriform plate. The *hypophysis* (pituitary gland) will also break off. Its point of attachment is the *infundibulum*. See Table 7 for a list of the cranial nerves and their functions.

Table 7
Cranial Nerves and Their Functions

Cranial Nerve		*Function*
I.	Olfactory	Sensory: smell
II.	Optic	Sensory: vision
III.	Oculomotor	Motor: movement of eye; also iris, lens, and eyelid
IV.	Trochlear	Motor: rotation of eye
V.	Trigeminal	Motor: tongue and jaw. Sensory: teeth and upper facial area
VI.	Abducens	Motor: rotation of eye
VII.	Facial	Motor: facial, jaw, neck. Sensory: taste
VIII.	Vestibulocochlear (Auditory)	Sensory: hearing and equilibrium
IX.	Glossopharyngeal	Motor: pharynx. Sensory: taste and touch
X.	Vagus	Motor: pharynx, vocal cords, and most internal organs
XI.	Spinal Accessory	Motor: pharynx, larynx, neck
XII.	Hypoglossal	Motor: tongue

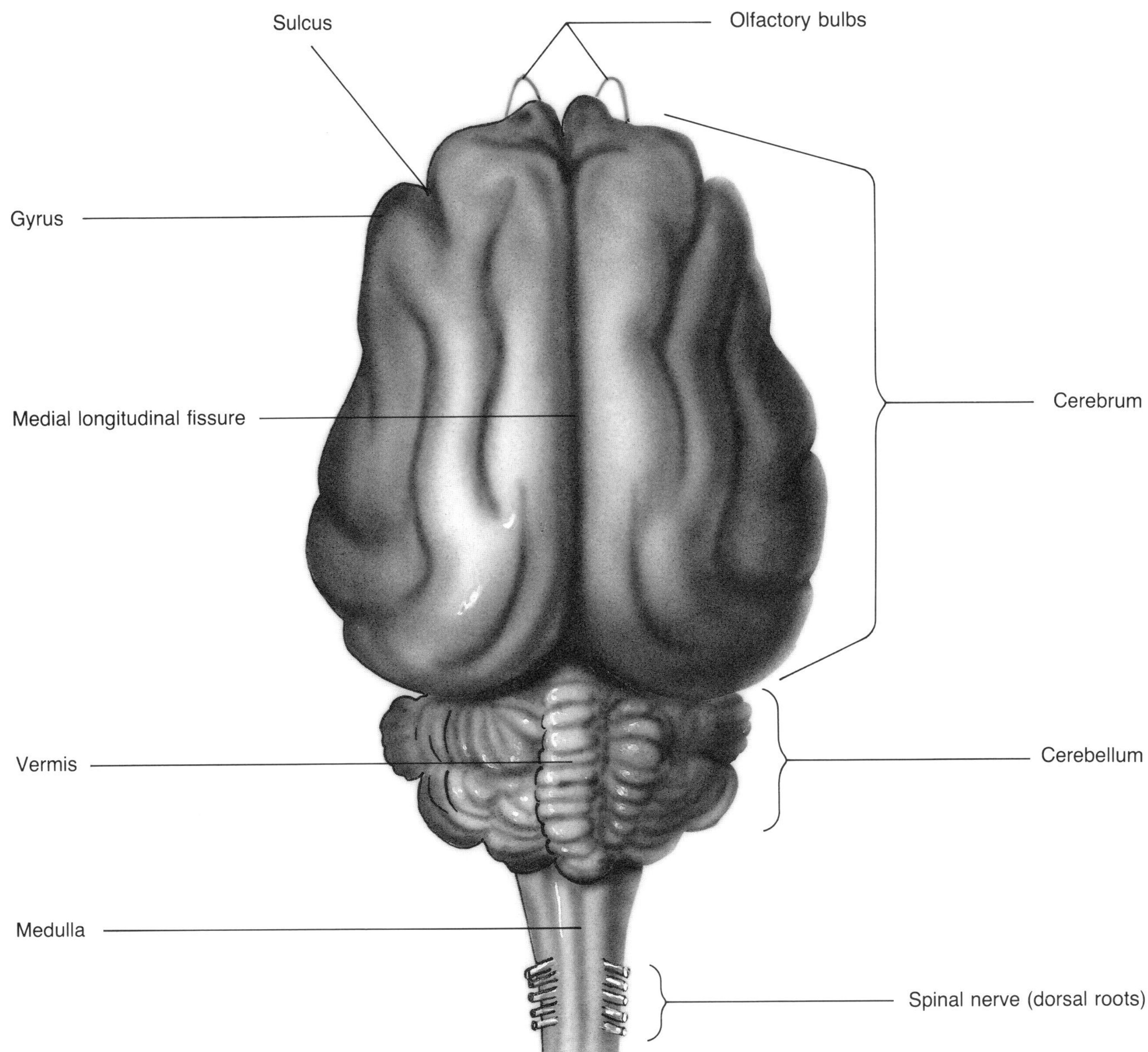

Figure 47. Brain (dorsal view).

Place the brain, dorsal surface up, on a dissecting tray. Use your fingers to gently widen the *medial longitudinal fissure*. Insert a knife blade or thin scalpel into the fissure, and with a single, smooth cut divide the brain into two longitudinal halves. See Figure 49 for identification of the parts.

Cerebrum. The cerebrum is the largest brain part and the seat of higher functions such as memory, learning, etc. There are left and right *cerebral hemispheres* divided by the medial longitudinal fissure.

The two hemispheres are connected by a broad band of nerve fibers, the *corpus callosum*. The *pyriform lobes* are separated from the rest of the cerebrum by the *rhinal fissure*. The pyriform lobes, *olfactory bulbs, olfactory tracts,* and other deeper parts form the *rhinencephalon,* the part of the brain concerned with smell.

The surface of the cerebrum is folded into a series of ridges or *gyri* (singular, gyrus). The furrows between gyri are *sulci* (singular, sulcus). These ridges and furrows increase the surface area of the cerebrum and therefore its gray matter.

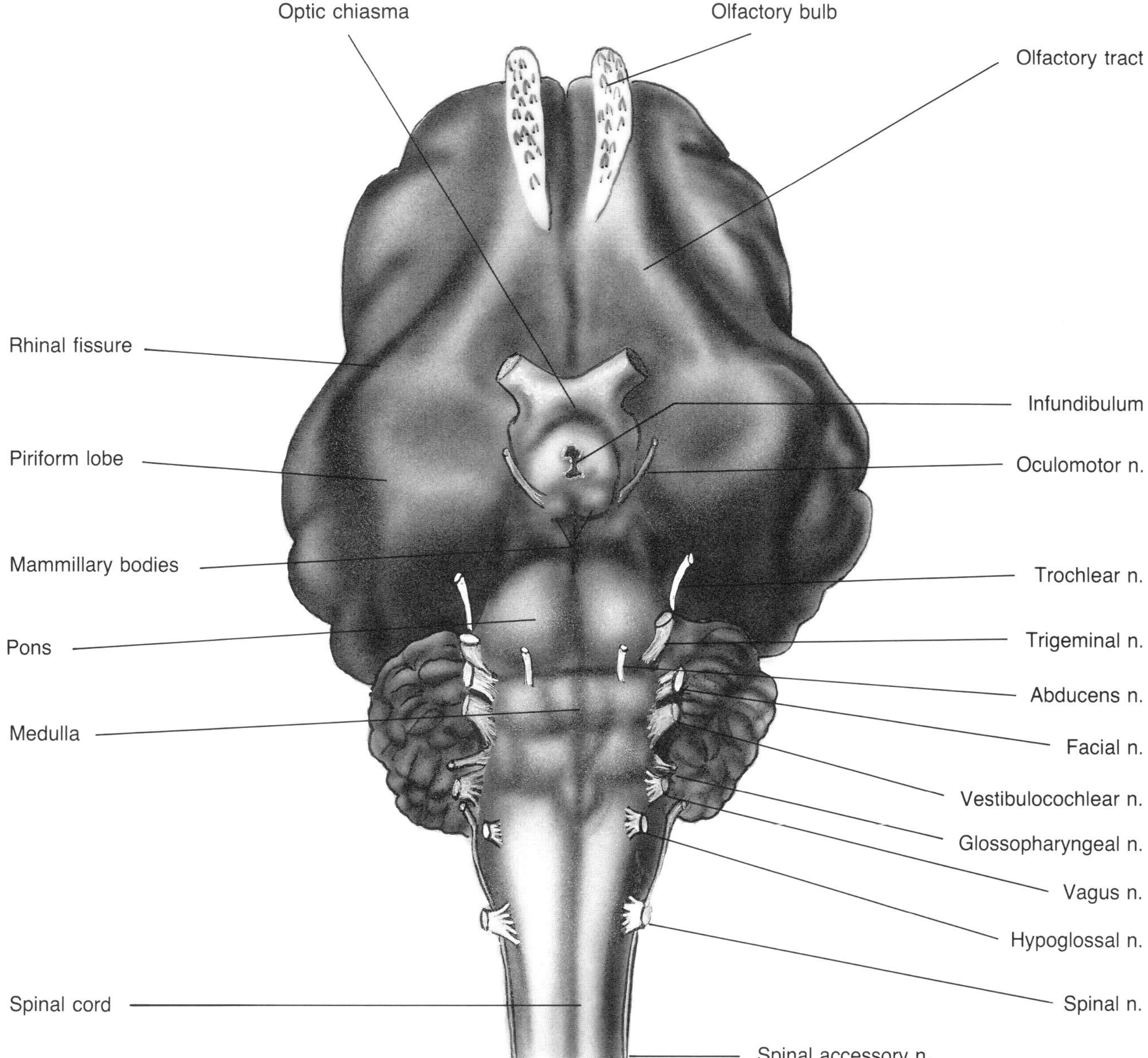

Figure 48. Brain (ventral view).

Cerebellum. The cerebellum is in many ways similar to the cerebrum. It also has an outer cortex of gray matter, is folded, and is divided into two hemispheres. The division is produced by a dorsal, central ridge, the *vermis*. The cerebellum is an important center for muscular coordination.

Medulla. The medulla is the most posterior portion of the brain. All nerve tracts between the spinal cord and higher parts of the brain pass through the medulla. The medulla also contains centers that regulate autonomic functions such as heart rate,

blood pressure, and respiration. The *pons* is an important center for nerve tracts that connect the cerebellum with other parts of the brain and spinal cord.

Tectum. The tectum is a part of the midbrain. It is a center for responses to visual and auditory stimuli.

Thalamus. All sensory nerves other than the olfactory nerves enter the thalamus. There, their impulses are sent to the appropriate higher centers for interpretation. The *hypothalamus* is directly ventral to the thalamus and regulates homeostatic processes

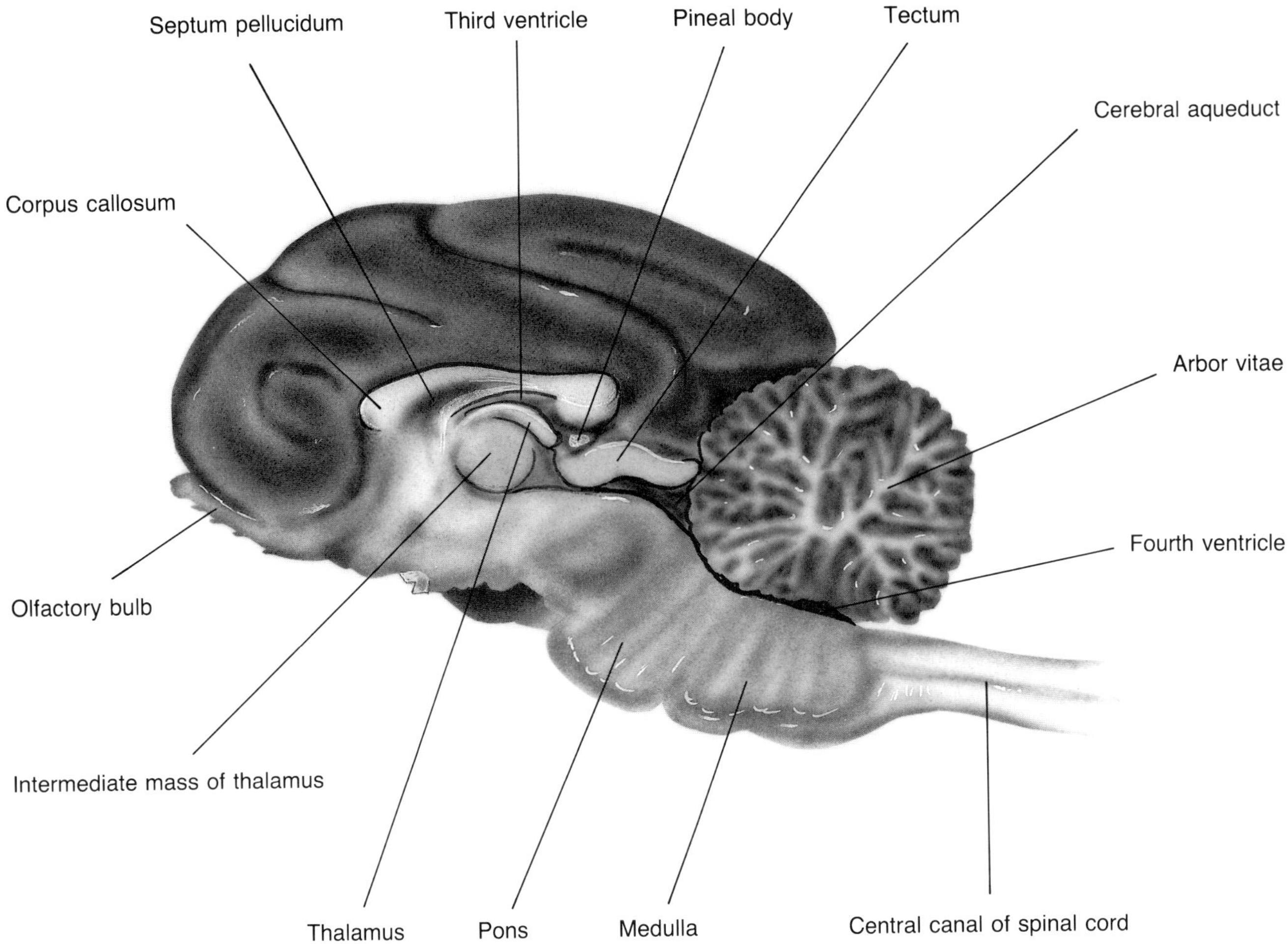

Figure 49. Brain (median view).

such as body temperature. It also controls the secretion of hormones by the pituitary gland.

Ventricles. The ventricles are four spaces within the brain. They communicate with the *central canal* of the spinal cord. In life they and the central canal contain the cerebrospinal fluid. The fourth ventricle is connected to the third ventricle by the *cerebral aqueduct.* The first and second (lateral) ventricles can be seen by removing the *septum pellucidum* in the transected brain.

Questions and Activities

1. Dissect the complete spinal cord.
2. Do a more complete dissection of a spinal nerve, exposing the dorsal and ventral roots.
3. If you have dissected the brains of other vertebrates, make a brief comparison of their brains to the cat's brain.

ADDITIONAL READING

The following reference books will be of aid both to the instructor and to the student of anatomy. Crouch's text has perhaps the most exhaustive set of illustrations of cat anatomy available and is highly recommended. The texts by Breland, Hyman, and Stromston are standard works on anatomy. Getty's text offers a different view, that of veterinary anatomy.

BRELAND, OSMOND P., *Manual of Comparative Anatomy*, McGraw-Hill, New York, 1943.

CROUCH, JAMES E., *Text-Atlas of Cat Anatomy*, Lea & Febiger, Philadelphia, 1969.

GETTY, ROBERT, *The Anatomy of the Domestic Animals*, 5th ed., W. B. Saunders Co., Philadelphia, 1975.

HYMAN, LIBBIE HENRIETTA, *Comparative Vertebrate Anatomy*, 3rd ed., The University of Chicago Press, Chicago, 1979.

STROMSTON, FRANK A., *Davison's Mammalian Anatomy*, 7th ed., The Blakiston Company, Philadelphia, 1937.

NOTES